ELECTROMAGNETIC MODELLING OF POWER ELECTRONIC CONVERTERS

THE KLUWER INTERNATIONAL SERIES
IN ENGINEERING AND COMPUTER SCIENCE

POWER ELECTRONICS AND POWER SYSTEMS

Consulting Editor

Thomas A. Lipo
University of Wisconsin - Madison

Other books in the series:

SPOT PRICING OF ELECTRICITY
Fred C. Schweppe
ISBN 0–89838–260–2

RELIABILITY ASSESSMENT OF LARGE ELECTRIC
POWER SYSTEMS
Roy Billinton and Ronald N. Allan
ISBN 0–89838–266–1

MODERN POWER SYSTEMS CONTROL AND
OPERATION
Atif S. Debs
ISBN 0–89838–265–3.

ELECTROMAGNETIC MODELLING OF POWER
ELECTRONIC CONVERTERS
J.A. Ferreira
ISBN 0–7923–9034–2

ENERGY FUNCTION ANALYSIS FOR POWER
SYSTEM STABILITY
M.A. Pai
ISBN 0–7923–9035–0

ELECTROMAGNETIC MODELLING OF POWER ELECTRONIC CONVERTERS

by

J.A. FERREIRA
Rand Afrikaans University

KLUWER ACADEMIC PUBLISHERS
BOSTON/DORDRECHT/LONDON

Distributors for North America:
Kluwer Academic Publishers
101 Philip Drive
Assinippi Park
Norwell, Massachusetts 02061 USA

Distributors for all other countries:
Kluwer Academic Publishers Group
Distribution Centre
Post Office Box 322
3300 AH Dordrecht, THE NETHERLANDS

Library of Congress Cataloging-in-Publication Data

Ferreira, J. A., 1958–
    Electromagnetic modelling of power electronic converters / by J.A.
Ferreira.
        p. cm. — (The Kluwer international series in engineering and
computer science Power electronics and power systems)
    Includes bibliographies and index.

    1. Electric current converters—Mathematical models. 2. Electric
current converters—Computer simulation. 3. Electromagnetism—
Mathematical models. 4. Electromagnetism—Computer simulation.
I. Title. II. Series: The Kluwer international series in
engineering and computer science. Power electronics & power
systems.
TK2796.F47   1989
621.381 '044—dc20   89–15501
                                                        CIP

ISBN 978-1-4419-5118-2

CONTENTS

Foreword

The era of the personal computer has, without doubt, permanently altered our life style in a myriad of ways. The "brain" of the personal computer is the microprocessor (together with RAM and ROM) which makes the decisions needed for the computer to perform in the desired manner. The microprocessor continues to evolve as increasingly complex tasks are required. While not sharing the limelight of the microprocessor, the "heart" of the personal computer, namely the power supply, is equally important since without the necessary source of power the microprocessor would be a useless piece of silicon. The power supply of twenty years ago was much different than its modern day equivalent. At the dawn of the personal computer era in the late 1970s, dc power was obtained from a simple diode bridge. However, the need for smooth, regulated DC at low voltage required at the same time both a bulky input transformer and a large dc side filter. Those computer fans present at the birth of this industry can remember the large boxes housing our Altair, Cromemco and Northstar computers which was made necessary largely because of the huge power supply. It is not well appreciated but certainly true that the huge sucess of the Apple II computer in those days was due, at least in part, to the relatively slim profile of the machine. This sleek appearance was largely due to the adoption of the then new and unproven *switched mode power supply*. The switched mode power supply uses power transistor technology to perform the necessary power conversion from AC to regulated DC. As a result the size of the transformer and filter can be drastically reduced.

Since a power supply is an inevitable component of all electronic equipment, the technology of switched mode power supplies continues to evolve and to increase in importance. In particular, the advent of the power MOSFETs has permitted switching at hundreds of kilohertz. Increased switching speeds are of vital importance to decrease filter components and lead lengths and thus improve efficiency and minimize cost. Research efforts are now being made to research the megahertz switching range. However, as circuit designers reach for ever higher switching speeds the basic assumption of classic theory, namely that the wavelength of all significant current components of the circuit must be larger than the circuit dimensions, begins to break down. The so-called "parasitic" effects such as skin and proximity effect, stray capacitance and losses of inductors and the like become of crucial importance in predicting circuit performance. In this book Dr. Ferreira presents a refreshing alternative to the use of circuit theory for analysis of high frequency switched mode converters, namely circuit design from an *electromagnetic* viewpoint. This book should go a long way towards the process of reorienting power supply engineers to different (but inherently more accurate) methods of analyzing high frequency switching circuits.

Thomas A. Lipo, Consulting Editor
University of Wisconsin-Madison

Preface

During the early 1980's electromagnetic design of power converters was identified by the Electronic Power Control Research Group at the Rand Afrikaans University to be an important aspect of future development in power converter technology. A motivation of this viewpoint was that the design of switchmode converters, operating at high frequencies, requires careful layout to contain stray impedances, while magnetic components are potentially more exposed to eddy currents, as embodied by the skin and proximity effects. Subsequently an investigation was conducted into the analysis of power electronic converters from an electromagnetic viewpoint. In short, this book is about a deeper understanding of the electromagnetic phenomenon of energy flow in conducting structures, develops suitable equations and algorithms to quantify the electromagnetic effects, and translates the theory into user friendly computer programs.

In the first part of the work, it is shown that qualitative analysis of electromagnetic energy flux, as depicted by the Poynting vector, gives an insight into the mechanism of power conditioning which is beyond the scope of classic circuit theory. Principles of energy flux transmission, when applied to switchmode circuits, give perspective to the origin of "stray" or "parasitic" impedances and eddy current losses in circuits.

An analytical method to calculate the resistance, inductance and capacitance of any combination of round, rectangular or plane conductors is described. It uses some aspects microstrip theory, in combination with the skin and proximity effect theory which forms the main theme of the book.

Circuit design with the objective of keeping the skin effect and proximity effect losses to a minimum is identified as a major aspect of the optimisation of magnetic components of high frequency converters. To this end, equations are derived to calculate ohmic losses in transformer and inductor windings. A prominent feature of this approach is that it permits separate calculation of skin effect, proximity effect and frequency components of eddy current losses. These equations are subjected to experimental verification, to support their validity.

An effective experimental method to measure the ac resistance of transformers and inductors was achieved by reviving a method developed during the first half of this century and subsequent developement to its full potential using modern measuring equipment. A feature of this method is that it relies to a great extent on the accuracy of the prediction model to make experimental observations.

A major objective of the work was also to create some tools for the electromagnetic design and optimisation of converters. To this end three CAD programs were developed which have been used to good effect by our laboratory and a few laboratories in West Germany. Since there do not seem to be anything comparable on the software market these programs are being made available for general usage*. Details of these programs can be found in Appendixes B to D of this book.

Braham Ferreira

*The three programs are available as a package with a manual included, and can be obtained from:

> J.A. Ferreira
> Energy Laboratory
> Rand Afrikaans University
> P.O. Box 524
> Johannesburg 2000
> South Africa

The programs are slightly upgraded versions of those described in this book. The programs run on MS[1] DOS personal computers equipped with a mathematical coprocessor and either a Hercules[2] monochrome or EGA colour graphics card. For non-profitable use such as university education, the programs are available at $200.00. When used gainfully for research and development, the price of the CAD software is $1000.00. Any payment should be made to "Rand Afrikaans University".

1) MS[TM] is a trademark of Microsoft Corporation
2) Hercules[TM] is a trademark of Hercules Computer Technology

ELECTROMAGNETIC MODELLING OF POWER ELECTRONIC CONVERTERS

CHAPTER 1

CIRCUIT ANALYSIS FROM AN ELECTROMAGNETIC VIEWPOINT; DESIGN OPTIMISATION BEYOND THE SCOPE OF CONVENTIONAL CIRCUIT ANALYSIS TECHNIQUES

Eddy currents give rise to frequency dependency of the impedance of conductors, transformers and coils. The effective resistance can easily become orders of magnitude larger than its dc value and can have a pronounced effect on the circuit behaviour and system performance. This introductory chapter gives a qualitive description, supported by examples, on this topical but often underrated phenomenon.

1.1 INTRODUCTION

The preference of alternating current to flow at the periphery of solid conductors has been known to electrical engineers for more than a century. In itself, this characteristic would not be a nuisance if it did not lead to extra loss of energy. Due to the losses and heating, the effective ac resistance appears to be larger than the actual dc resistance and because of the implications on the system performance, it has been receiving due attention during the design of 50/60 Hz transformers, inductors, machines and high frequency transmission lines. Especially at high current levels or high frequencies this effect calls for special construction techniques of conductors. Not only the resistance is affected, but the internal inductance also becomes a function of frequency, a factor that must be reckoned with in many circuits. However, due to the complicated nature of the current redistribution, the analytical solution becomes impractical except for very simple cases. Before the advent of the modern computer, theoreticians were only able to give guidelines as to the optimum conductor configuration and manufacturers had to rely to a great extent on empirical data.

The availability of power semiconductors capable of fast switching and accurate control of power, and magnetic materials with excellent high frequency properties, resulted in power converters operating in the kilohertz and even megahertz frequency range. The magnetic components are consequently not only subjected to high frequencies but also to non-sinusoidal waveforms. While the scope of application became larger and covers a bigger frequency range, the author has experienced that many engineers involved in this field may not observe the correct general design philosophy of coil and winding design. Even in the international literature on power electronics, few papers mention the design and optimisation of magnetic components in the circuit and how it influences the efficiency of the circuit.

1.2 SKIN AND PROXIMITY EFFECTS

Current flowing in a conductor gives rise to a magnetic field inside and outside the conductor and causes the familiar skin effect. One explanation often put forward is that the inductance of a central filament of the conductor is higher than for peripheral filaments, since the flux linkage is higher in the centre of the conductor. The longitudinal voltage drop inside the conductor must be constant over the cross section, and because the impedance on the inside is larger than on the periphery, most current flows on the outside. Consequently, compared to the voltage drop along the conductor, the current at the periphery is leading while the core current is lagging.

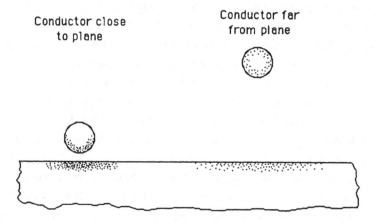

Figure 1.1 Skin effect and proximity effect in conductors.

A conductor subjected to skin effect is considered to be isolated from any external magnetic field sources. The redistribution of current towards the surface occurs only because of the magnetic field generated by its own current. This assumption is generally not valid in circuits, for the current is perturbed by magnetic fields of neighbouring conductors due to the proximity effect. The mutual influence of current densities in forward and return conductors, which carry current in opposite directions is portrayed by the example in figure 1, where the one conductor is a wire and the other, (as commonly encountered in power converters,) a current carrying plane, which could be an aluminium heatsink. Note that the current concentrates itself on the surfaces closest to the other conductor which correspond with the sections of the conductor cross section, where the inductance and therefore also impedance is the smallest.

From a circuit viewpoint the skin and proximity effect causes frequency dependence of the circuit impedance. The effective ac resistance is related to the depth of penetration which is inversely proportional to $\sqrt{f}$, giving a resistance to frequency relationship that approaches a $\sqrt{f}$-slope at frequencies where the skin depth is much smaller than the conductor dimensions. The effective inductance of a conductor, coil or winding is made up of two terms: the first is the external component due to the external magnetic field, and the second is the actual internal inductance resulting from the internal field. The internal inductance is a function of the current distribution in the interior; since this heterogeneous distribution consists of an increase in the density at the periphery, the internal term decreases. The skin and proximity effects therefore decrease the effective inductance as a function of frequency. As an example, the resistance and inductance of the configuration in Figure 1.1 have been plotted in Figures 1.2 and 1.3, utilising 2 mm copper wire, a 10 mm thick aluminium plane and plane to conductor-distances of 1 mm and 10 mm. Note that the resistance of the closely spaced conductors is almost double that of the widely spaced case. Also, in the closely spaced case, the frequency dependence of the inductance is more pronounced because the internal inductance dominates the external inductance.

In closely spaced conductors, the proximity effect is most pronounced in the outermost conductors. To illustrate this, consider a case of five parallel conductors, far removed from any other magnetic field source and all carrying the same current, as shown in Figure 1.4. In the centre conductor the skin effect dominates, in the second conductor the current is concentrated on the outer surface of the conductor, while in the outermost conductors phase reversal of the current on the inner and outer surfaces is possible.

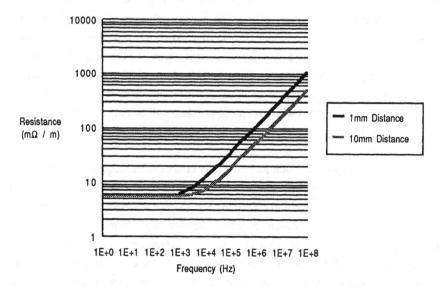

Figure 1.2 Effective per meter length resistance of wire/plane current path.

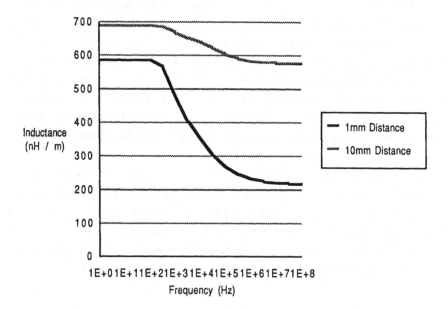

Figure 1.3 Effective per meter length inductance of wire/plane current path.

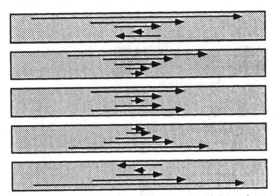

Figure 1.4 Proximity effect in five parallel conductors.

1.3 INDUCTOR COILS

Specifying and winding an inductor of a given size is usually straight forward with the aid of an inductance bridge. Obtaining the maximum possible Q-factor or, more applicable to power conversion, the highest reactive power rating per cost unit, involves the highest inductance and lowest ac resistance for the shortest and thinnest conductor. It is interesting to note that the tendency has never been to employ optimal solenoid cross section winding configurations, such as the Brooke inductor, which gives the most inductance for a given length of wire, in applications with a high frequency content in the current such as choke coils in power converters and high Q coils in telecommunication equipment. This is because the proximity effect tends to be the dominant conductor loss mechanism; arranging the windings to obtain the shortest magnetic path lengths and high field intensities inevitably enhances the proximity effect.

To illustrate the influence of the coil configuration on the effective ac resistance a series of coil designs as illustrated in Figure 1.5 are compared to each other. All the coils, with the exception of the coil with a magnetic core, are designed for an inductance of 104 μH and they all have equal average diameters. The copper cross section areas of all the conductor types utilised are the same apart from the litz wire case where the cross section area is somewhat less due to the insulation between the strands. In case of wire wound coils the diameter is 2mm, while the copper strip conductor has thickness of 0.5mm and width of 6.28mm. In the litz-wire instance, the number of strands are 10, each with diameter of 0.6mm. The ac resistance graphs in the subsequent figures have been computed with the algorithms and equations described in this book. Despite the occurrence of some prediction error, the experimental work supports the general shape of the graphs.

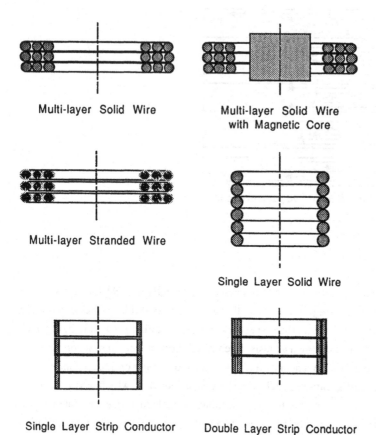

Figure 1.5 Cross section of coil configurations.

Figure 1.6 compares the ac-resistance curves of a 5 layer, 5 turns per layer coil and a single layer 33 turn coil wound with the same wire and both rated for the same inductance value. The multilayer design uses less wire and has therefore a lower dc-resistance, whereas the proximity effect is less pronounced in the single layer coil, which performs better at high frequencies.

In the quest for low ac resistance, one finds that the resistance does not necessarily improve with thicker wire. Keeping everything the same while increasing the wire diameter, the ac resistance decreases only up to a certain point, after which the slope becomes positive (see the example in Figure 1.7). This phenomenon can become a serious limitation if the objective is to obtain low ac resistance at high frequencies, since the minimum resistance diameter is related to the depth of penetration at the frequency of interest. Further improvements in reducing the losses can be obtained by stranding

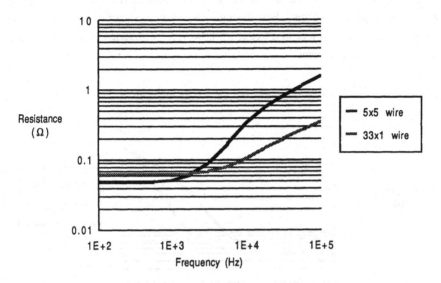

Figure 1.6 Ac resistance of single layer versus multilayer coils.

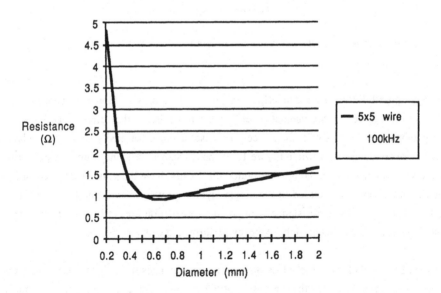

Figure 1.7 Variation of resistance with wire diameter at 100 kHz.

of the conductor. Individual wires of the strand are enamelled and weaved along the entire divided conductor in such a way that all wires successively pass through all points of the cross-section, which ensures that the current will divide equally among the seperate strands. In Figure 1.8 is shown, how the high frequency resistance can be improved by using stranded wire. At a certain frequency however, the stranded conductor becomes inferior to a single solid wire.

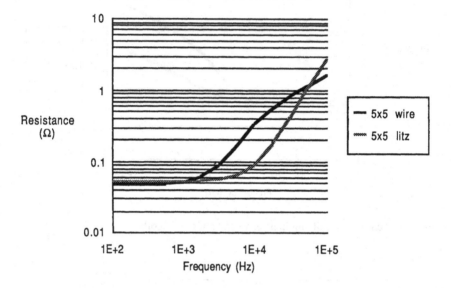

Figure 1.8 Comparison between air cored inductors with solid wire and stranded (litz) wire.

Another possibility is to use thin strip conductors instead of wire. However, care should be taken to ensure that the magnetic field lines run parallel to the strip surface because perpendicular field lines will cause excessive losses and one can easily be worse off than with wire conductors. Shown in Figure 1.9 is the resistance frequency curve of two strip conductor coils, which, although configured differently, have the same inductance value as the other coils. Compared to the single layer wire wound coil, one finds that it is difficult to match its dc resistance with equal cross section area strip, but on the other hand one can achieve better high frequency performance with strips.

Winding an inductor on a low loss magnetic core enhances the inductance for a given number of turns. Provided that the losses of the magnetic material is sufficiently small, one can generally achieve a better Q for large inductance values by utilising a core. Yet, the mere act of inserting a magnetic core inside an air cored inductor, perturbes the

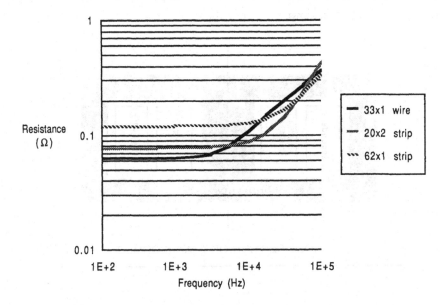

Figure 1.9 Ac resistance of coils wound with strip conductors.

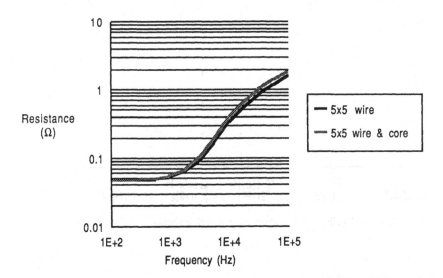

Figure 1.10 Comparison of the resistance curves of air core and magnetic core inductors with the same dimensions.

magnetic field surrounding the windings and therefore also influences the proximity effect and ac resistance. As an example a magnetic rod is inserted in a cylindrical coil which increases the inductance from 104 μH to 358 μH, but also results in a slight increase in ac resistance, shown in Figure 1.10.

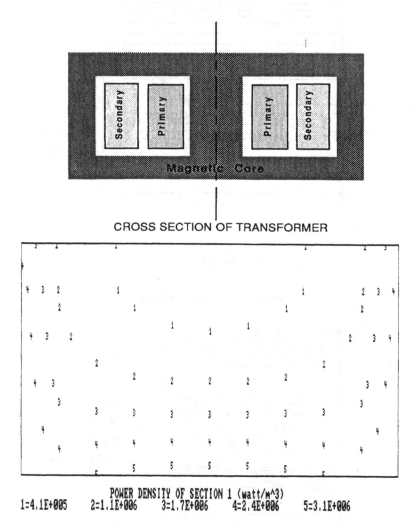

Figure 1.11 Loss distribution in the primary winding section of the transformer.

1.4 TRANSFORMERS

Transformer windings are essentially subjected to the same limitations as inductors, due to the proximity effect. The main source of magnetic field which causes proximity effect losses is not the magnetisation current, but the high intensity leakage field due to the transformer action. These losses are not evenly distributed across the windings, and the most losses occur at the interface between the primary and secondary windings. The loss distribution of the primary winding of the transformer used as example analised in this section is shown in Figure 1.11.

An effective way of reducing conduction losses is to subdivide the primary and secondary into a number of sections and then to interleave these sections. Because the primary and secondary currents oppose each other, this method effectively reduces the magnitude of the magnetic stray field which again reduces losses. Figure 1.12 shows the improvement which can be achieved in terms of the primary referred resistance if the primary is divided into three sections and the secondary into two.

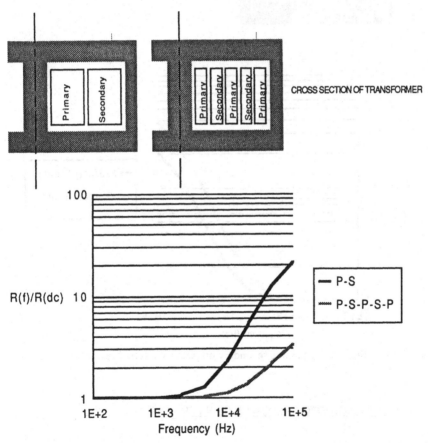

Figure 1.12 Primary referred resistance of a sectionalised and unsectionalised transformer.

Something which is not often appreciated is that losses do occur in passive secondary windings of multi-secondary transformers. To illustrate this effect a transformer with two secondaries, as shown in Figure 1.13, is analysed. In the first case the secondary closest to the primary is connected to a load and in the second instance the outermost secondary is connected.

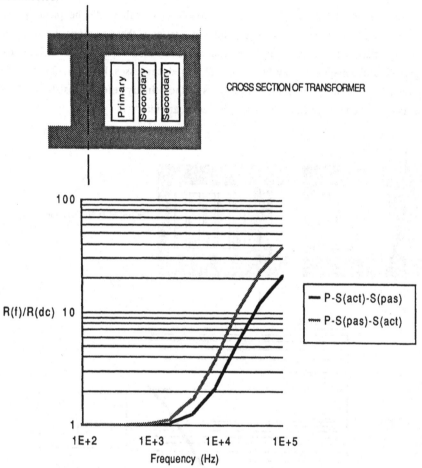

Figure 1.13   Influence of a passive secondary on primary referred resistance.

1.5 ELECTROMAGNETICS IN POWER CONVERTERS

When laying out the circuit of a switch mode converter, it is important to contain the value of structural and parasitic impedances. Measured current and voltage waveforms of switching transients in many cases have damped oscillations on them which are caused by parasitic and structural impedances in the circuit. These impedances have their origin in the physical geometry of the active and passive components and interconnections of the circuit and, due to its nature, is very much an electromagnetic field problem. The amplitude and energy content of the oscillations can pose a threat to the electronic switches and it is possible to reach a point where the reactive power, due to the structural inductance and capacitance, could influence the efficiency of the circuit.

As illustrated by the examples of this chapter, the most obvious electromagnetic effect is the losses due to eddy currents. The combined skin and proximity effect has been shown to influence the effective ac resistance of magnetic devices, sometimes in an unexpected manner. Circuit analysis shows that a great portion of converter losses occurs in inductors and transformers. Power converters which feed electrical machines, impose current waveforms on these machines with frequency components at which the ac resistance might be orders of magnitude larger than the dc resistance of the armature. It is therefore important for the design of a power converter, whether it be a 50 W switch mode power supply operating at 1MHz or a 1 MW industrial drive at 50Hz, to take note of the losses as it has a great effect on the obtainable ratings of the system and the cost per watt output power.

The ceaseless pressure on the power-frequency product frontier of switch mode power converter technology, is pushing the ability of conventional circuit analysis techniques to optimise such a system to its limits. Concerning the theory, one finds that the modelling techniques are entering the grey area between classic circuit theory and classic electromagnetic theory. This book contains a study of electromagnetic modelling techniques with the purpose to develop analytical tools which, in conjunction with with classic circuit analysis techniques, creates the possibility to achieve higher degrees of circuit optimisation. In addition, a new approach on the concept of energy flow and its relation to skin and proximity effects is presented.

CHAPTER 2

POYNTING VECTOR, A METHOD TO DESCRIBE THE
MECHANISM OF POWER CONDITIONING

Qualitive analysis of electromagnetic energy flux as depicted by the Poynting vector, gives insight into the mechanism of power conditioning which is beyond the scope of classic circuit theory. It is contemplated that an understanding and eventually an analytical grip on energy flux and its interaction with matter is essential to optimise energy conversion at high switching frequencies.

2.1 HISTORICAL PERSPECTIVES

Electromagnetic theory provides two approaches of viewing an electromagnetic phenomenon. The most commonly used approach analyses field lines or currents and voltages, and is the backbone of present day modelling techniques. The other, equally valid method, views energy flux or, as it is being presented in textbooks, the Poynting vector. Oliver Heaviside[13] summarised the analogy between the two methods appropiately as follows:

> "Now in Maxwell's theory there is the potential energy of the displacement produced in the dielectric parts by the electric force, and there is a kinetic or magnetic energy of the magnetic induction due to the magnetic force in all parts of the field, including the conducting parts. They are supposed to be set up by the current in the wire. We reverse this; the current in the wire is set up by the energy transmitted through the medium around it....."

The rate of storage and dissipation of energy inside a volume element can be expressed in terms of energy flux vector **S**, commonly known as the Poynting vector **S**:

$$\int_A \mathbf{S}.d\mathbf{A} = -\partial\left(\int_v (\varepsilon E^2/2 + \mu H^2/2)\ dv\right)/\partial t - \int_v \mathbf{J}.\mathbf{E}\ dv \qquad (2.1)$$

where $S = E \times H$ (W/m^2)   (2.2)

and $E \equiv$ electric field intensity (V/m)
        $H \equiv$ magnetic field intensity (A/m)
        $J \equiv$ current density (A/m^2)
        $\varepsilon \equiv$ permittivity (F/m)
        $\mu \equiv$ permeability (H/m)

It is customary to use the Poynting vector to represent rates of flow of energy and momentum in electromagnetic waves. In electrical engineering it is often being used to analyse electromagnetic radiation and is also a useful technique for calculating eddy current losses. In circuit analysis however it has found limited application. On transmission lines it serves to illustrate "for educational purposes" that power flows in the space surrounding a conductor and not to our preconception that it should be inside the conductor. The application of the Poynting vector to describe electromagnetic power flow inside idealised electrical machines has been advocated by a few academics[3,4,7,8,10], because it provides an alternative approach to gain insight into its operation . Despite their efforts, the Poynting vector concept in electrical machines has gained very little support, and MMF-field energy relation treatment is generally being accepted as the best method of analysis.

The reason for the apparent unpopularity of the Poynting vector lies deeper than merely difficult vector analysis. Doubts existed concerning the physical significance of energy flowing in closed paths accompanied by zero divergence of S which can therefore not be observed. An example of this phenomenon would be a permanent magnet placed in a static electric field which creates a circulating energy flux.

Since S is defined only in terms of its divergence, an infinite number of functions, $S = E \times H + C$ where $\text{div}(C) = 0$ satisfies this definition. J. Slepian suggested various alternative Poynting vectors with their associated energy densities[2]. One worth mentioning is obtained by setting:

$$C = \nabla \times VH$$ (2.3)

and $$E = -\partial A/\partial t - \nabla V$$ (2.4)

When C and E are substituted in equation (2.2) we get:

$$S = S + C = VJ + V\,\partial D/\partial t + (H \times \partial A/\partial t)$$ (2.5)

The following significance can be attached to each term in equation (2.5);

$\mathbf{V}\mathbf{J} \equiv$ Longitudinal flow of energy

$\mathbf{V}\partial\mathbf{D}/\partial t \equiv$ Radial flow of energy to be stored in the electric field

$\mathbf{H} \times \partial\mathbf{A}/\partial t \equiv$ Radial flow of energy to be stored in the magnetic field.

The above variation of the Poynting vector, also called the Slepian vector, contradicts the Poynting postulate that the energy flux is merely guided by conductors while energy transfer occurs in the surrounding space. The VJ-term confines the steady state energy flow to the conductor material. Such apparent contradictory view points of where the energy flux really flows definitely did harm to the acceptibility of a theory which describes energy flow. Another factor that counted against the Poynting vector was that it was not linked properly to the laws of energy and momentum conservation.

As recently as 1964, however it was shown that Poynting vector momentum is necessary to avoid violation of the law of momentum conservation[5,6]. This puts the Poynting vector on a sounder footing and excludes alternative energy flux vectors, such as the Slepian vector, from electromagnetic theory[9]. This discovery of the "hidden momentum" of the Poynting vector and subsequent follow up papers has almost completely been confined to physics journals. As a matter of fact, it seems to have gone almost unnoticed by the engineering community, and with a few exceptions, such as textbook[11], none of the work, published on the Poynting vector since 1964, mention this new development in energy flux analysis.

The Slepian vector needs however, not be discarded altogether. As long as one keeps in mind that it does not contain the full energy flux picture, it can still be used. Its strong point is that it analyses the two types of field energy conversion in an elegant manner, provided the scalar and vector potential references are carefully chosen.

2.2 THE ENERGY FLUX VIEWPOINT OF POWER ELECTRONICS

In circuit theory the tendency is to consider parasitic effects such as stray inductances and eddy currents to be a nuisance which are either assumed to be small enough to be ignored or are included as an afterthought to explain measured waveforms, which do not behave as expected. Up to now the tendency has been to handle parasitics in an ad hoc manner, and it seems to the present author that misconceptions concerning stray impedances exist among many practising engineers in this field. The present-day emphasis on high frequencies in switching converters puts the parasitic effects more in the limelight, and the ability to optimise their role in power electronic circuits is becoming a crucial requirement. The purpose of this chapter is to advance energy flux analysis as being a method to get a grip on these parasitic impedances. It must however be emphasised that it is not envisaged that energy flux as an analytical method can compete with circuit analysis, which is deeply entrenched in electrical engineering practice. The Poynting vector would rather serve to give the circuit engineer a qualitative feeling of what is good practice from an energy transmission viewpoint.

This chapter confines itself to energy transmission in the macroscopic sense in power circuitry. In other words, we restrict ourselves to classic electromagnetics and detail of how the Poynting vector is converted from or into alternative energy forms, as in batteries, semiconductors, resistive, magnetic and dielectric materials, is beyond the scope of this treatise since it involves analysis at molecular level, which would inevitably include quantum effects.

The subsequent paragraphs of this chapter illustrate some of the "laws" governing energy flow in electric circuits. It is presented in a qualitative way, deliberately steering clear of electric and magnetic field describing functions, which can generally be found in electromagnetic textbooks. Poynting vector energy flow is schematically represented by solid black lines in diagrams. Energy flux which is transmitted over a section of line is terminated by an arrow, while energy absorbed within the section ends in a dot.

2.3 PROPAGATION OF A PULSE ALONG A TRANSMISSION LINE

Low ohmic connections between devices in circuits form the backbone of all electric circuitry. Any connection comprises of a forward and return path for the current and therefore requires two leads. At high frequencies such an interconnection is better known as a transmission line because current and voltage signals can be observed in terms of their propagation along the line. Consider for example the two wire parallel transmission line depicted in Figure 2.1. The surface over which the Poynting vector is being integrated is cylindrically shaped, as shown in the figure. By making the diameter of the cylindrical surface large, the Poynting vector becomes zero on axial surface 3 due to a decrease in value of at least inverse proportionality to the square of the radial distance, and the fact that electromagnetic propagation along the line is in the TEM-mode. From equation (2.1) then, the nett gain in field energy and ohmic losses inside the cylinder can be obtained by integrating S over the two surfaces 1 and 2.

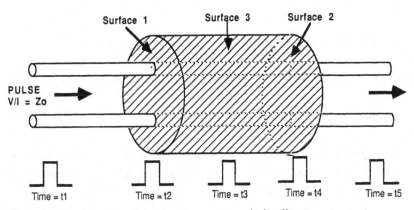

Figure 2.1 Propagation of a pulse along a transmission line

Let a voltage and current pulse (at a ratio determined by the characteristic impedance of the line) propagate along the line at the speed of light. If one could freeze the charge distribution along the line at a specific time instant, no nett charge will be present on the line except in the section occupied by the pulse. Accompaning the charge distribution is a current inside the conductor and electric and magnetic field intensities in the space surrounding the pulse section. At time t_1 the pulse has not yet reached the cylinder and all the quantities in equation (2.1) are zero. At time t_2 the pulse enters the cylinder and the surface integral of ExH over surface 1 is exactly equal to the product VI, which in turn is equal to the rate of increase in electric and magnetic field energy inside the cylinder and the I^2R ohmic losses inside the conductor leads enclosed by the cylinder. When the pulse is inside the cylinder (time t_3) the surface component over the cylinder equals zero, no power leaves the cylinder, while a small fraction of the field energy is converted into ohmic losses. At t_4 the pulse leaves the cylinder, impressing a non-zero ExH only on surface 2, and the surface integral will be equal to the decrease in field energy inside the cylinder together with ohmic losses. After the pulse has left the cylinder, everything will be in electric equilibrium once again, and a very slight increase in the temperature of the conductors being the only trace of the transition of a pulse.

2.4 POYNTING VECTOR IN POWER LINES

Applying Maxwells equations to the transmission line problem, the following expressions in terms of voltage and current can be derived:

$$\partial V/\partial z = -RI - L\partial I/\partial t \qquad\qquad (2.6)$$

$\partial I/\partial z = -GV - C\partial V/\partial t$ (2.7)

where V,I,R,L,G,C are normal circuit symbols in per unit length quantities.

If equation (2.6) is multiplied by I, (2.7) by V, and the results added, we obtain[12]:

$\partial(VI)/\partial z + \partial(^1/_2LI^2 + ^1/_2CV^2)/\partial t + (RI^2 + GV^2) = 0$ (2.8)

which is nothing else than the Poynting vector equation in terms of circuit analysis symbols, written in differential form. The first term gives the change in power along the length of the transmission line, which can either be converted into field energy (the second term) or dissipated (third term).

If the waveforms are sinusoidal we can utilise the complex phasor notation, writing Z = R + jX and Y = G + jB, in which case (2.6) and (2.7) become:

$dV/dz = -(R + jX)I$ (2.9)

$dI/dz = -(G + jB)V$ (2.10)

Multiplying (2.9) by I^* and adding to it the product of V and the conjugate of the second equation, gives;

$d(^1/_2VI^*)/dz + ^1/_2(RII^* + GVV^*) + j/_2(XII^* - BVV^*) = 0$ (2.11)

In power electronic circuitry the line lengths are generally much shorter than the wavelength of significant frequency components in current and voltage waveforms. Under this assumption the analysis reduces to quasi-steady state. Analysing the power line in the frequency domain, equation (2.11) would be applicable. The imaginary part gives the rate of change of "reactive" power. It can be zero under certain conditions, in which case the line is free of reactance. The real part on the other hand gives the rate of change of average power flow and the average power loss in resistance and conductance per unit length.

Subsequently we proceed to describe the energy flow in power lines, in an attempt to give equation (2.11) meaning in terms of the Poynting vector. Firstly consider an ideal power line as illustrated in Figure 2.2. The conductors are lossless and the load is matched, i.e. the ratio, voltage to current, on the lines is equal to the characteristic impedance of the conductor configuration. The conductors are assumed to be wide strips, resulting in the Poynting vector being confined to the space between the lines. All the energy flux entering the one side of the line is available on the other side; none of it gets lost along the way, is dissipated or (temporarily) stored in parasitic reactances. In the case of an ideal power line the input and output current and voltage values are identical at any instant (allowing for propagation time).

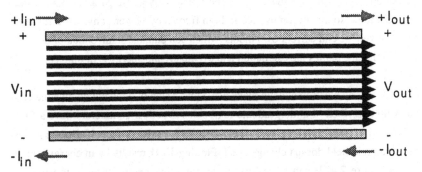

Figure 2.2 Poynting vector energy flux along a lossless matched line.

In the second case study as illustrated in Figure 2.3, the conductors are lossless, but the voltage to current ratio is not equal to the characteric impedance of the line. If the ratio is larger, the line will appear to be capacitive, and the output current will either be smaller or larger than the input value, depending whether the stray capacitance is being charged or discharged. For a smaller ratio the line will appear inductive and the output voltage will increase or decrease depending on whether the inductance is storing or delivering energy.

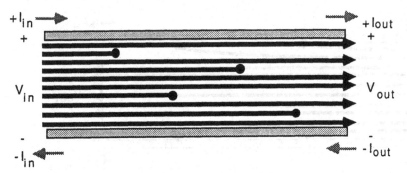

Figure 2.3 Energy flux along a lossless mismatched line.

Thus, theoretically a connection between two devices can be free of stray reactance, provided the spacing between the forward and return conductors is selected to match the current and voltage levels. Electromagnetic energy can be transferred unimpeded in an electric circuit at a ratio of electric to magnetic field intensity determined by the characteristic impedance of the medium. Through integration of the electric and magnetic field in and around the conductors, the value of current along and voltage between the conductors can be calculated, the ratio of which gives the characteristic impedance of the line. Any deviation from this ratio will cause deposition of field energy along the line, thus reducing the power delivered to the load. The energy stored along the line is recoverable and, depending on the circuit conditions, may cause parasitic oscillations which is a common source of annoyance in high frequency power converter circuits. It tends to cause voltage and current overshoots, which can damage semiconductor devices. Resistance in the line damps the oscillation and eventually the energy is dissipated.

The penetration of an electromagnetic plane wave into a slab of good conductive material is extensively described in undergraduate textbooks. Practical conductors have a finite conductivity which causes a longitudinal electric field on the surface of the conductors while the magnetic field doesn't change at all. Plotting **ExH** results in an energy flux as depicted in Figure 2.4. It can be seen that the Poynting vector curls in towards the conductors, causing ohmic losses in them. Despite the fact that conductive material converts electromagnetic energy flux very efficiently into heat, a small fraction of the surrounding energy flux crosses the interface between the air (free space) and the conductor, due to the big difference in the **E** to **H** ratio of energy flux in the two media, which causes most of the incident energy flux to be reflected at the air conductor boundary. In case of superconductors energy flux will not enter the conductive material , which gives rise to the viewpoint that conductors merely serves to guide TEM waves, preventing them to disperse.

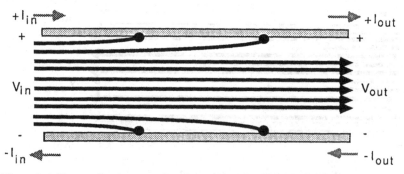

Figure 2.4 Energy flux in a matched line and leads with finite conductivity.

2.5 FLOW OF ENERGY AROUND OBSTACLES

Next consider the case where a conductive cylinder is put between the two conductors, as despicted in Figure 2.5. What happens is that the energy flux is deflected around the obstacle, resulting in higher flux density between the cylinder and conductor. It doesn't matter whether it is an ac or dc signal on the line. Even if the cylinder is made of a good dielectric or magnetic material the resulting flux lines will be the same. The underlying cause for this deflection of the Poynting vector flux is not a mystery, hidden in Maxwells equations; the reason lies in the fact that the propagation velocity decreases the better a conductor, dielectric or magnetic material one has. The energy flux then simply follows the path of least "resistance" and travels around the obstacle.

Figure 2.5 Deflection of energy around a cylinder.

Positioning a conductive obstacle in an "ac" energy flux causes eddy currents to be induced in order to curve the energy flux, which in turn results in ohmic losses in the obstacle. In case of the obstacle being made of dielectric or magnetic material the energy flux interacts with the molecules, changing the energy level of the material. Since penetration of the energy flux into the material involves energy conversion, it experiences resistance which deflects the flow around the object, towards the path of easy transmission. Viewing a power line connection in this manner can be useful because obstacles made of practical materials can be lossy. By tracing the energy flux paths in a power converter one can for instance identify obstacles such as bolts which inadvertently lie in the path of high levels of energy flux and give cause to unwanted losses.

Transmission line theory would describe the effect of such an obstacle in terms of impedance mismatch and reflected waves. Let us assume for a moment the incident energy injected from the left (Fig 2.5) is a step function with a very fast rise time. The moment the energy front hits the obstacle, only part would be transmitted while the remainder would be reflected. The reflected energy front returns to the source where it could either be absorbed or once again be reflected. The consequence of this to and fro propagation of energy flow is two fold; firstly, energy is dammed up on the left and the effective rate of energy transfer is retarded. The "energy dams" can contain either magnetic field energy or electric field energy depending on the material type of the obstruction. By dividing the energy flow into a series of small pulses, one can visualise each pulse being reflected up and down in the section left of the obstacle (assuming a mismatch at the source), and each time it hits the obstacle a fraction of the energy is transmitted. The time it takes for an "energy packet" to get past the obstacle can then directly be related to the number of up and down trips it takes to transmit all its energy past the obstacle. Utilising this phenomena is general practice in electric circuits; a low pass LC-filter is nothing else than a means to retard energy propagation along the circuit!

2.6 POYNTING VECTOR IN CAPACITORS

In circuit theory a capacitance is defined in terms of a differential equation $I = CdV/dt$, and its energy relation $W = \frac{1}{2}CV^2$. Transmission line theory on the other hand gives verdict on the structure of a capacitor and the distribution of the charging current. The capacitor dimensions, are normally smaller than the wavelengths associated with the frequencies encountered in power converters. During capacitor construction a dielectric is inserted to enhance the capacitance and to improve its energy storage capacity. Viewing a capacitor from an energy flux viewpoint provides an apparent paradox; namely, that dielectric material is better at reflecting energy than storing it. For example, if a power line is terminated with either an ideal conductor (infinite conductivity) or an ideal dielectric (infinite permittivity) the outcome will be the same; the line will be shorted, any energy flux will be fully reflected and the line will appear inductive at its input.

Consider the capacitor structure in Figure 2.6. The end is open circuited and a slab of dielectric is placed between the parallel conductors, leaving two small air-gaps between the dielectricum and conductors. Assuming the length of the capacitor to be much smaller than the wavelength value, the current would decrease linearly from a maximum at the entry point to zero at the open end. Because the voltage between the conductors stay essentially constant (assuming high conductivity), a linearly decreasing longitudinal Poynting vector ($E \times H$) exists along the length of the capacitor.

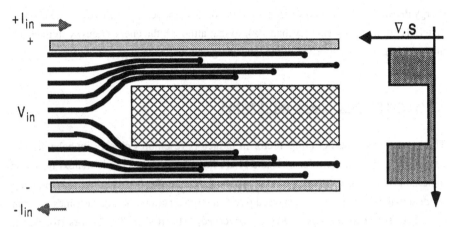

Figure 2.6 Energy flow in a capacitor structure.

The linear decrease in longitudinal energy flux indicates that the flux is either diverted to the transverse direction or is being converted into another form. One would expect the latter to take place because a capacitor stores energy in an electric field. For a good dielectric the electric field, E, inside the material will be very small. Therefore the expression for static energy density, $1/2 \mathbf{D}.\mathbf{E}$, assigns the biggest energy density to the space between the dielectric and conductors and a smaller value to the dielectric itself due to the small value of $\mathbf{E}$. The term $V\partial \mathbf{D}/\partial t$ from equation (2.5) describes the energy flow associated with electric field energy conversion. To avoid ambiguities that can arise due to the selected reference point of the electric scalar potential V, $\nabla.\mathbf{S} = \mathbf{E}.\partial \mathbf{D}/\partial t$, which gives energy conversion, is rather used (and plotted in Figure 2.6).

The previous example might seem to contradict the presupposition that the dielectric slab will act as a sink of recoverable energy. Even though the dielectric does show an inclination to reflect energy in this example, its insertion enhances the total structure's energy storage capability because it enforces a higher electric field intensity for the same applied voltage in the air gaps while the dielectric also have a larger energy capacity for a given electric field intensity. The electric field energy inside a gab then is effectively increased because the density is determined by the square of the field intensity. On the other hand, if no airgap exists between the dielectric and conductors, the electric field, E will be the same as before the slab was inserted and the energy density $1/2\mathbf{E}.\mathbf{D}$ will increase by a factor equal to the relative permeability. In this case energy is stored in the dielectric slab, because the energy flux is given no alternative path past the dielectric. This argument might seem to lead to inconsistencies when one assumes infinite permeability and infinitely small gaps, since it leads to infinte values of electric field intensity and energy density. One must however bear in mind, that practical permittivities of dielectric materials are normally smaller than 100, so that although the energy density in the gaps might be larger, the proportion of the total energy stored in the gaps will be small.

2.7 POYNTING VECTOR IN INDUCTORS

The behaviour of the Poynting vector in inductors resembles closely its behaviour in capacitors, provided the roles of $\mathbf{E}$ and $\mathbf{H}$ are interchanged. In circuit context magnetic material also acts as a reflector of energy flux. For example, by inserting a (toroid) ring of ideal magnetic material in a shorted power interconnection, the line becomes an open circuit. In so far as their geometry is concerned, capacitors and inductors (for power converter applications) differ completely. Because the electric field lies in the same plane as the current flow, the capacitor structure becomes an extension of a transmission line structure. The fact that the magnetic field is perpendicular to the plane of current flow, gives rise to the magnetic structures also being placed perpendicularly. Due to the closed loop nature of the magnetic fields, the magnetic structure becomes a magnetic circuit analogous to an electric circuit. In this respect the energy flux method offers an advantage; it combines the electric and magnetic circuits into a quantity which is topical in power circuits, namely energy flow.

Consider the air cored helical inductor shown in Fig. 2.7. Assume that a source located at the left delivers energy to the coil. The energy flux from the source is given by $\mathbf{H}_1 \times \nabla V$, where V is the applied voltage and $\mathbf{H}_1$ the magnetic field due to the line current. At the inductor surface, two electric field components can be identified: firstly there is the axial electric field between the winding which is simply an extension of the applied field $-\nabla V$. The other component is the induced component, of which the integral around the circumference of the coil is predicted by Faraday's law. This electric field intensity is directed tangentially to the curvature of the coil and can be expressed as $-\partial A/\partial t$ with $\mathbf{A}$ being the magnetic vector potential. The magnetic field $\mathbf{H}_2$ is associated with energy storage and is directed parallel or anti-parallel to the axis of the coil. The cross product of storage magnetic field and induced electric field, $\mathbf{H}_2 \times \partial A/\partial t$ gives rise to an energy flux which is radially directed away from the coil surfaces, and which is responsible for the charging of the magnetic field $\mathbf{H}_2$. Discharging of the inductor is accompanied by a reversal in the induced field and an energy flux directed radially in towards the coil conductors.

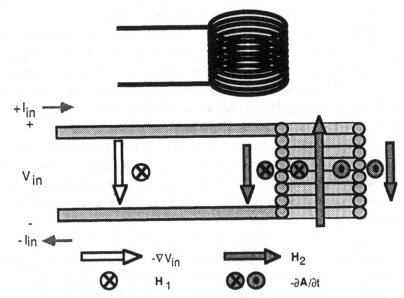

Figure 2.7 A coil and the Poynting vector constituents indicated on its cross section.

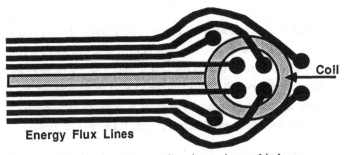

Figure 2.8 Top view of energy flow in an air cored inductor.

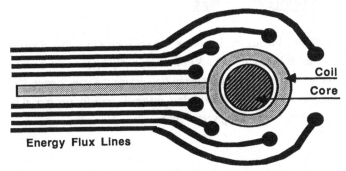

Figure 2.9 Top view of energy flow in an inductor with a magnetic core.

Fig. 2.8 pictures the energy flux associated with an air cored inductor. Energy flux is converted into field energy both on the inside and outside of the coil. If a magnetic core is placed in the centre of the coil, the magnetic field intensity decreases on the inside while at the same time increases in magnitude on the outside (Fig. 2.9). Because the energy density varies with the square of the magnetic field, more magnetic energy storage per ampere line current is obtained by inserting a magnetic core. The energy flux also pictures the change in energy storage location by indicating that the inward energy flow is substantially reduced while an increase in energy flow directed radially outward can be detected. As a matter of fact, in all cases one is able to trace the energy flow towards the region of field energy storage. For example, in an inductor wound on a closed core with an airgap, the fringing of the magnetic field round the gap serves to converge the energy flux into the airgap.

As one would expect, the value of the Poynting vector decreases the deeper the medium of magnetic storage is penetrated. Taking the cylindrical volume enclosed by the air coil for example, H_2 tends to be reasonably constant over the radial cross section, while A decreases almost linearly towards the centre where it becomes zero.

Studying the energy flux patterns in inductors is not of academic interest alone, but is a very useful tool in optimising an inductor design. The eddy current losses in both the coil and magnetic material is closely related to the energy flux intensity at any point in the inductor. (For more details see Chapter 4). Since the Poynting vector in an inductor, once understood, responds closely to our preconceptions to what is "natural"; it provides a method of evaluating a structure without having to resort to detailed numerical analysis.

2.8 POYNTING VECTOR IN TRANSFORMERS

Figure 2.10 shows a transformer linking an alternating voltage source to a load. Energy flow along the conductor leads is described by VI or alternatively by the vector cross product, $H \times \nabla V$. Not so obvious however is the energy flow inside the transformer. No "physical" voltage or current is present and the standard magnetic transformer model is not of much assistance in finding the Poynting vector either.

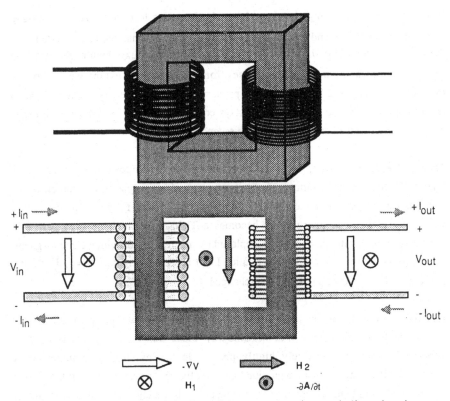

Figure 2.10 A transformer and the Poynting vector constituents indicated on its cross section.

The operation of a transformer can be split in two parts; namely an inductor which involves energy exchange between the primary and core, and secondly the transformer mechanism of energy exchange between the primary and secondary windings[1]. Usually a designer strives to keep the energy exchanges between the primary and core, or in circuit terminology; the magnetisation current as small as possible. Increasing the permeability effectively reduces the energy flow to the core and in the limit confines energy flow only to the transformer mechanism.

The electric field component active in the winding window is the magnetic potential induced component $-\partial A/\partial t$. Its direction is tangential to the pitch of the coils. The active magnetic field is none other than the stray field, indicated by H_2 in Fig. 2.10. The stray field is linked to the MMF distribution, due to transformer operation, over the length of the winding window. Between the primary winding and core, its value is zero and it increases to a value of $N_p I_p/(\text{window width})$ in the space between the primary and secondary. While crossing the secondary, H_2 decreases to become zero in the gap between the secondary and the core leg. The transformer mechanism is associated with the Poynting vector $H_2 \times \partial A/\partial t$, which describes an energy flux which builds itself up from zero to a maximum as one progresses from left to right inside the window of the transformer in Figure 2.10, crossing the primary winding. The energy flux is directed from the primary to the secondary and stays constant untill the secondary winding is crossed where it is then channelled into the secondary circuit.

The transformer window appears to be a cavity that traps electromagnetic energy flux injected by the primary winding. The magnetic core effectively acts as a screen which encloses the energy flux in the window. If the windings extend fully over the permitted width of the window, energy flux becomes a TEM wave, and the core acts as a transmission line, with the conductors replaced by magnetic leads which carry magnetic flux instead of current. Instead of the relation $I = \int H.dl$ one now has $V = \int E.dl$, and the roles of $E(= -\partial A/\partial t)$ and H are interchanged.

Exactly the same arguments holds for a magnetic "transmission line" as for the electric power line discussed in par. 2.4. In the same way "ideal" transmission can be achieved if superconductors and infinitely highly permeable lossless magnetic material are utilised,and if this magnetic transmission line is configured in such a way that the electric and magnetic field energy between the magnetic leads are equal. In this way a lossless transformer can be constructed which is free of stray parasitic impedances. In practical transformer structures, due to limitations on magnetic materials and operating frequency, the stray capacitance due to the $\partial A/\partial t$ field is much smaller than the leakage inductance due to the H_2 field in Figure 2.10, with the result that the latter dominates. The effect of inter turn capacitance is also a factor to be taken into consideration and normally dominates the $-\partial A/\partial t$ capacitance.

Electromagnetic modelling is not merely of academic interest but has direct bearing on the efficiency of the transformer. In a multilayer winding each layer contributes in an equal amount to the power transmitted or received by that winding. It results in the energy flux steadily building up until it reaches a maximum at the primary-secondary boundry. The energy flux, as it crosses winding conductors, causes eddy current losses in them which is a main source of losses in high frequency power transformers.

2.9 POYNTING VECTOR IN ELECTRIC MACHINES

Energy flow in electric machines, viewed by means of the Poynting vector is covered quite extensively in literature, (see references [1,3,4,7,8,10]) and no need exists to repeat it all here. Only a brief discussion is therefore presented.

A machine slot can be considered to be one half of a transformer; in case of a motor, the primary side is applicable and for a generator the secondary. The magnetisation of the machine can be described in terms of the inductor model discussed in paragraph 2.7. Two Poynting vector components responsible for magnetisation can be identified: one which points towards the magnetic material of the machine and another large component directed towards the air-gap.

The construction of the machine is done in such a way that a periodic distribution of the magnetic vector potential is created along the circumference in the air-gap when the machine is energised. When this A-distribution is being rotated relative to the armature, whether it being by mechanical or electrical means, an electric field $-\partial A/\partial t$ is created in the machine slots. Loading a machine is always associated with an increase in armature current. This current, which excludes the magnetisation component, sets up a tangential magnetic field H_t in the airgap. The cross product $H_t \times \partial A/\partial t$ which lies in a radial direction, supplies the Poynting vector which effectively couples a mechanical source/load to an electrical load/source.

The conversion of $H_t \times \partial A/\partial t$ into voltage and current is similar to the transformer action described in the previous paragraph. The Poynting vector, $H_t \times \partial A/\partial t$, is really nothing else than an intermediate step toward mechanical power conversion. To get the mechanical power we first have to convert A in terms of the magnetic "force" field B:

$$B = \nabla \times A \qquad\qquad (2.12)$$

Assuming a cylindrical rotor of a highly permeable material, yields a magnetic flux normal to the rotor surface, radially directed towards or away from the rotor axis, i.e.;

$$B = B_n r \qquad\qquad (2.13)$$

The tangential component of mechanical stress (N/m^2) at a point on the surface of the rotor is conveniently given by the product of the normal flux density and the tangential field component H_t at that point;

$$S_S = B_n H_t \qquad (2.14)$$

The mechanical torque density is simply the stress devided by the rotor radius r

$$t_s = B_n H_t / r \qquad (2.15)$$

If the rotor rotates at ω the power density becomes:

$$p_m = B_n H_T \omega / r \qquad (2.16)$$

The power density given by equation (2.16) is applicable to all kinds of machines. Energy flux of the form $H \times \nabla V$ enters the electric ports of the machine via the power lines and is converted into $H \times \partial A/\partial t$ by the coils of the machine. The $H \times \partial A/\partial t$ format of the Poynting vector is generated to attain two objectives. Firstly the energy flux is decoupled from the conductor leads and is able to traverse the air gap distance without the aid of power lines. This Poynting vector can in the second place easily be converted into mechanical (Maxwell) stresses which is responsible for the mechanical energy conversion.

2.10 REFERENCES.

[1] J Slepian; "The flow of Power in Electrical Mchines"; *The Electric Journal;* Vol XVI ; Pt 5; pp 303-311, 1919.

[2] J Slepian; "Energy and Energy Flow in The Electromagnetic Field; *Journal of Applied Physics*; Volume 13, August1942; pp512-8.

[3] EI Hawthorne; "Flow of Energy in DC Machines"; *AIEE Transactions*; Vol 72; Pt 1 ; pp 438-444; September 1953.

[4] EI Hawthorne; "Flow of Energy in Synchronous Machines"; *AIEE Transactions*; Vol 73; Pt 1, pp 1-9; March 1954.

[5] RP Feynman, RB Leighton and ML Sands; "The Feynman Leectures on Physics"; Vol II; pp 17-6 and 27-11; Addison-Wesley, Reading, Mass.,1964.

[6] EM Pugh, GE Pugh; "Physical Significance of the Poynting Vector in Static Fields"; *Amer. J. Phys.;* **35**:153(1967).

[7] BB Palit; "Unified Analysis of Electric Machines with the help of Poynting Vector and the Electromagnetic Energy Flow in the Air-gap space. Part I: General Theory"; *Zeitschrift fuer angewandte Mathematik und Physik;* **31**, pp 384-399; 1980.

[8] BB Palit; "Unified Analysis of Electric Machines with the help of Poynting Vector and the Electromagnetic Energy Flow in the Air-gap space. Part II: Application of the General Theory"; *Zeitschrift fuer angewandte Mathematik und Physik,***31** ; pp 400-412; 1980.

[9] PC Peters; "Objections to an alternate Energy Flow Vector"; *Amer. J. Phys.*; Vol. 50 ; No. 12; December 1982.

[10] CB Cray; "Air-gap Power Flow and Torque Developement in Electric Machines - can we teach the Fundamentals?"; *IEEE Trans. Power Apparatus & Systems*; Vol PAS-103; No. 4; April 1984.

[11] A Shadowitz; The Electromagnetic Field, McGraw-Hill Kogakusha, Ltd., 1975; pp421-429.

[12] S Ramo, JR Whinnery, T van Duzer; "Fields and Waves in Communication Electronics";John Wiley & Sons,Inc; 1965.

[13] O Heaviside; "Electrical Papers Vol 1"; Macmillan, p438, 1892.

[-] This chapter apeared in edited form as a paper:
    JA Ferreira; "Application of the Poynting Vector for Power Conditioning and Conversion"; IEEE Trans. Education; Vol. 31, No 4, November 1988, pp 257-264.

CHAPTER 3

POWER CONDITIONING IN ELECTRONIC CIRCUITS

Principles of energy flux transmission, as described in the previous chapter, are applied to switchmode electronic circuits. An investigation is conducted into the role of "stray" and "parasitic" impedances during power conditioning at high frequencies, as means to gain insight into the optimisation of circuits operating at high power-frequency ratings.

3.1 INTRODUCTION

Traditionally in low frequency power electronics the conductive leads, which serve to interconnect devices in a circuit, are viewed to be "ideal" with no resistance, inductance or capacitance. Increasing the frequency or repetition rate of the current and voltage signals, causes the reactance or capacitive leakage to increase directly proportional to frequency, while at the same time the resistance rises steeply due to skin and proximity effect. One then finds that, at a certain point, the stray lead impedances begin to influence the circuit waveforms and it becomes necessary to include the effect of the stray impedance on the circuit behaviour. The stray impedances are mostly considered a nuisance because it tends to hamper and limit the circuit operation, in particular parasitic inductance of transistor leads and snubber connections. If he had an understanding of these parasitic effects, an engineer should, as with many other applications in the engineering field, succeed to turn the apparent problems into advantages by utilising the parasitic impedances to improve the waveforms.

Modelling of magnetic components normally involves an equivalent circuit which includes a magnetising inductance, a series resistance and leakage inductance. At higher frequencies the leakage inductance and stray capacitance, as with connection structures, start playing a more prominant role, and because magnetic components become smaller, the thermal time constant is smaller, which makes high frequency transformers and inductors more vulnerable to overloads. Another factor which is often neglected is that the resistance and leakage inductance is strongly frequency dependent and special care should be taken during the design of magnetic components. Later in this book techniques are developed to calculate and predict impedances, which entail the main body of this work. In this and the previous chapter an energy flux approach is adopted to gain insight into role the "parasitic" and "stray" effects play in power conditioning of converters.

This chapter is a continuation of Chapter 2, as the principles previously descibed are being applied to electronic circuits. A connection structure is introduced, which is essentially a piece of short transmission line, (compared to the wave lengths of the frequencies associated with the waveforms,) having a geometry that is applicable to power electronic circuits. In the first part of the chapter the connection structure is modelled in terms of well known circuit networks, which, with the aid of algorithms and theory being developed in the other chapters of this thesis, can be calculated and applied. The second part of this chapter is devoted to the introduction of a concept that uses concatenation of connection structure elements to do energy flux based analysis.

3.2 ENERGY FLOW IN A SWITCHMODE CIRCUIT EXAMPLE

To conclude this paper energy flow in the well known flyback converter is discussed. The flyback converter circuit, shown in Figure 3.1, includes a mutual inductor with an air gapped core, as shown in the energy flow diagrams. During the first part of the cycle the transistor in the primary circuit conducts, and the current in the inductor steadily increases, while the diode prevents any current from flowing in the secondary circuit. When the transistor is turned off, the inductor current is transferred to the secondary winding, and current flows via the diode to the load. Capacitors are included in the circuit to smooth out the large ripple current associated with this type of switchmode converter.

The energy flow in the flyback converter is illustrated in Figure 3.1 in a schematic manner. To simplify the presentation, the conduction losses in the leads, windings and magnetic core have not been included, and the magnetic material is assumed to be infinitely permeable so that all the magnetic energy is stored in the air gap. The capacitors are parallel plate structures capable of storing electric field energy, while the connection between the capacitor and inductor, which includes the transistor, operates at a V/I ratio smaller than its characteristic impedance, so that it acts as a parasitic inductance to the circuit. Futhermore the input and output energy flux is assumed to be constant.

CIRCUIT DIAGRAM

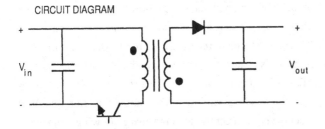

ENERGY FLOW DIAGRAM BEFORE TRANSISTOR TURN-OFF

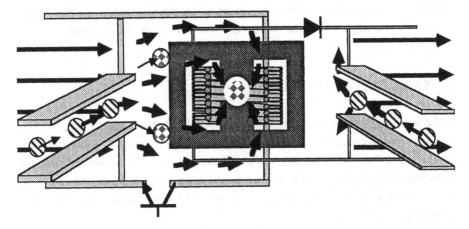

ENERGY FLOW DIAGRAM IMMEDIATELY AFTER TRANSISTOR TURN-OFF

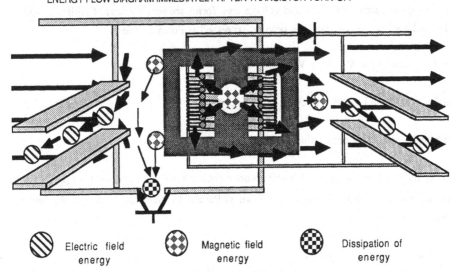

Electric field energy   Magnetic field energy   Dissipation of energy

Figure 3.1 Circuit diagram and energy flow in a flyback converter

During the conduction cycle of the transistor, both the energy flow from the capacitor and incident energy flux is directed to the air gap of the inductor. Part of this energy is however absorbed by the parasitic inductance of the primary circuit, which at turn-off is directed to the transistor where it is dissipated. Note that while the transistor is on, both capacitors supply power, while during the off-time they absorb power. If one makes more detailed energy flux inside the winding window of the mutual inductor, one finds that charging of the inductor occurs very efficiently, with the energy flowing unimpeded to the airgap. However, during discharge of the inductor, the energy flux crosses the primary winding on its way to the secondary circuit. This, as can be verified through numeric analysis programs, causes eddy current losses in the primary winding, which reduces the efficiency of the circuit during the cycle when the secondary conducts.

3.3 POWER TRANSFER EFFICIENCY

In the analysis of networks, there are two generally accepted approaches, namely, viewing of power-energy relations or determining of signal waveforms. In this paragraph the characteristics of a connection structure in terms of its power transfer efficiency is investigated. By viewing it as a two port that absorbs complex power, one is able to make an assessment of its contribution to the system efficiency and is also able to define impedance quantities.

Let us assume that the current and voltage waveforms are periodic and of arbitrary shape. The voltage and current can then both be expressed as Fourier series:

$$V(t) = \text{Real} \left(\mathbf{V}_0 + \mathbf{V}_1\, e^{j\omega t} + \mathbf{V}_2\, e^{j2\omega t} + ... + \mathbf{V}_n e^{jn\omega t} + ... \right) \qquad (3.1)$$

$$I(t) = \text{Real}\left(\mathbf{I}_0 + \mathbf{I}_1\, e^{j\omega t} + \mathbf{I}_2\, e^{j2\omega t} + ... + \mathbf{I}_n e^{jn\omega t} + ... \right) \qquad (3.2)$$

where $\mathbf{V}_n$ and $\mathbf{I}_n$ are peak value phasors.

Since the complex exponential functions are orthogonal over the interval t_0 to $(t_0 + T)$, we obtain the relationship, commonly known as Parseval's theorem, for the total complex power.

$$S = \sum_{n=0}^{\infty} S_n$$

$$= \sum_{n=0}^{\infty} V_n I_n^* \qquad (3.3)$$

Even at the highest harmonics encountered in the current and voltage waveforms the length of circuit interconnections are generally much shorter than the associated wavelengths of the frequency components of the signal. Therefore the signal amplitude will vary, for all practical purposes, linearly between the start and termination points of the interconnection.

By rewriting equation (2.11) in terms of the input and output complex power, we obtain the complex power loss per frequency component over the connection length:

$$\delta S_n = V_{n\text{-in}} I_{n\text{-in}}^* - V_{n\text{-out}} I_{n\text{-out}}^*$$

$$= \frac{(R_n I_{n\text{-av}} I_{n\text{-av}}^*)}{2} + \frac{j\omega(L_n I_{n\text{-av}} I_{n\text{-av}}^* - C_n V_{n\text{-av}} V_{n\text{-av}}^*)}{2} \qquad (3.4)$$

$$\text{where } I_{n\text{-av}} = \frac{(I_{n\text{-in}} + I_{n\text{-out}})}{2} \qquad (3.5)$$

$$V_{n\text{-av}} = \frac{(V_{n\text{-in}} + V_{n\text{-out}})}{2} \qquad (3.6)$$

and R, L, C = the resistance, inductance and capacitance of the line.

The total complex power loss is then:

$$\delta S = \sum_{n=0}^{\infty} \delta S_n. \qquad (3.7)$$

The reactive power loss over an interconnection structure can be written from (3.4) as follows:

$$\delta P_X = \frac{\sum_{n=0}^{\infty} (L_n I_{n\text{-}av} I_{n\text{-}av}{}^* - C_n V_{n\text{-}av} V_{n\text{-}av}{}^*)}{2} \tag{3.8}$$

The meaning of the above equation (3.8) is not unique. Zero reactive power as being understood in circuit theory does not necessary correspond with $\delta P_X = 0$, as can easily be verified. The reactive power will only be zero if each of the different frequency components are zero. Ideal power transmission through a connection structure can only take place if both current and voltage waveforms have exactly the same shape and if the V/I ratio is the same as the characteristic impedance of the line.

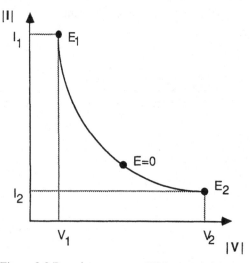

Figure 3.2 Reactive power on VI-locus

In practice one very seldom finds that the waveforms satisfy the conditions for ideal power transmission. If the ratio of the time domain current and voltage response of a connection structure in a switchmode converter is taken one would typically find that the reactive energy content alternates between states E_1 and E_2, as characterised by (V_1, I_1) and (V_2, I_2) on Figure 3.2. At state 1 the energy is stored in the magnetic field, the connection being inductive, and the reactive energy decreases until a point on the VI-locus is reached where its value is zero and the energy flow conditions are ideal at that instant. Beyond this point energy is being absorbed from the system by the connection and is stored in the electric field, as the locus moves to state E_2 causing the structure to appear capacitive.

3.4 MODELLING OF CONNECTION STRUCTURES IN CIRCUITS

As it will be indicated in later chapters, inductance and resistance are frequency dependent, thus complicating the summation of (3.7) somewhat. The biggest difficulties, however are encountered during simulation of circuit behaviour, because a frequency dependent impedance implies that the analysis is also to be conducted in the frequency domain which requires "advance" information of the time domain response. This inevitably adds to considerable complications and it would be useful to define equivalent impedance values which can be used to approximate the effect the frequency dependence has on the circuit behaviour. These equivalent values also require beforehand knowledge of the frequency components of typical waveforms, but in most cases it can be estimated from measurements or first order analyses of the circuit behaviour with sufficient accuracy.

The first step is to find the rms values of the waveforms on a conductor:

$$I_{rms} = \frac{1}{T} \int_{to}^{to+T} I^2(t)dt \qquad (3.9)$$

$$V_{rms} = \frac{1}{T} \int_{to}^{to+T} V^2(t)dt \qquad (3.10)$$

Equivalent impedances are then defined, based on the time domain and frequency domain power relations, as follows:

$$R_{eq} = \frac{\sum_{n=0}^{\infty} R_n I_{n-av} I_{n-av}^*}{2 I_{rms}^2} \qquad (3.11)$$

$$L_{eq} = \frac{\sum_{n=0}^{\infty} L_n I_{n-av} I_{n-av}^*}{2 I_{rms}^2} \qquad (3.12)$$

$$C_{eq} = \frac{\sum_{n=0}^{\infty} C_n V_{n-av} V_{n-av}{}^*}{2 V_{rms}{}^2} \qquad (3.13)$$

Above equations give equivalent impedance values of a connection structure, calculated from the frequency dependent impedance values in frequency domain, that can be used in time domain based simulations. Important of this approach is that it requires advance information of the Fourier components of typical voltage and current waveforms.

A connection structure may act predominantly inductive or capacitive depending on whether δP_X, from equation (3.8), is significantly positive or negative. In many such cases the connection need only to be modelled as being a stray inductance or capacitance; which is then defined as effective values as follows:

$$L_{eff} = \frac{\delta P_X}{I_{rms}{}^2} \qquad (P_{r-eff} > 0) \qquad (3.14)$$

$$C_{eff} = \frac{-\delta P_X}{V_{rms}{}^2} \qquad (P_{r-eff} < 0) \qquad (3.15)$$

3.5 CIRCUIT REPRESENTATION OF CONNECTION STRUCTURES

It is important that connection structures be adequately incorporated in the analytical model or numerical simulation of the circuit operation. As is indicated in the previous paragraph, the frequency dependence of the impedances can complicate the analysis considerably, even to the extent where analysis becomes impractical or impossible. In this paragraph equivalent circuits are presented, which can be incorporated in circuit analysis to predict voltage and current waveforms.

As indicated by Wheeler [2]; the frequency variation of the internal impedance (R and L) of a conductor can be analysed in terms of electric circuit elements as shown in Figure 3.3. It involves a hypothetical transmission line, structured in a manner and composed of enough elements to trace the impedance to frequency curve with satisfactory precision. However, such transmission line representations can be very time consuming in a simulation because of the many elements involved. Utilising T and Π-networks as it is being done in microstrip theory [1] (see figure 3.4) provides a simpler alternative. A T-network would normally be used when the capacitive term dominates, whereas for a mainly inductive line a Π-network can be utilised. The component-value are those given by equations (3.11 to 13). The representation can even be further simplified if the effective reactance value as presented in (3.14) and (3.15) are used. (see Figure 3.5).

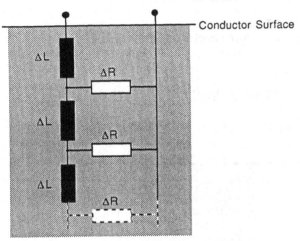

Figure 3.3 Network to simulate frequency dependence of internal resistance and inductance

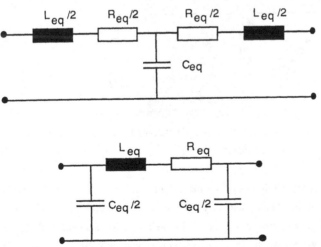

Figure 3.4 T and Π network representation of connection structures

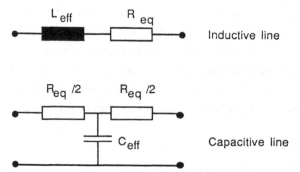

Figure 3.5 Connection structure presentation in terms of effective reactances

3.6 INTERACTION BETWEEN CONNECTION STRUCTURES AND ELECTRONIC SWITCHES.

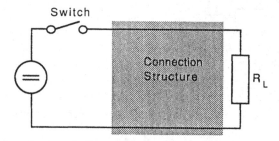

a. Switchmode control of power from a voltage source.

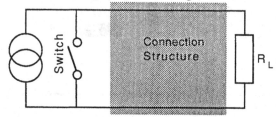

b. Switchmode control of power from a current source.

Figure 3.6 Two simple power electronic circuits

Figure 3.6 shows two circuit topologies of simple switchmode converters. The sources are connected and disconnected at a certain duty cycle to the load, thereby controlling the power flow delivered. For the purpose of simplifying the examples, a switch and source are assumed to be integrated, with the only connection structure being the section between the switch and the load. The losses in the connection are assumed to be negligible, thus resulting in a real valued characteristic impedance.

An energy flux approach to the operation of these circuits, requires the various elements to be viewed in terms of their characteristic impedances. A voltage source can be regarded as a very large capacitor and a current source as a very large inductor which have characteristic impedances that approach zero and infinity respectively. One generally finds that a source under normal circuit conditions never delivers power at its characteristic V/I ratio (impedance) and experiences the connection structure as a mismatched load.

We now consider the influence the impedance of the connection structure has on the waveforms and the switching stresses on the switches.

a) $R_L = R_o$:

The ideal case is when $R_L = R_o$. The switch experience no stress because it experiences a purely resistive load. Since no reflection occurs at the load the voltage and current waveforms changes shape as fast as the switch can operate.

b) $R_L < R_o$:

This is the most commonly encountered connection in switchmode circuits. The line appears inductive, and due to this the switches are stressed at turn-off, which compels the inclusion of turn-off snubbers. Turn-on is not stressed, but the rise time of the current in the connection and load is governed by the RL time constant, consisting of R_L and the connection structure. The rise time can be explained in terms of an energy flux which is repeatedly reflected between the load and source (Figure 3.6a) or switch (Figure 3.6b), while building up the additional magnetic field around the line, above the value required by the Poynting vector, to reach the steady state current as required by R_L.

c) $R_L > R_o$:

The connection structure appears to be capacitive and this causes turn-on stress in the switch due to the charging current peak. Turn-off is stress free, but the current and voltage has exponential waveforms due to the RC time constant. As in the previous case the waveforms can also be related in terms of energy flux, which is fully reflected at the switch (Figure 3.6a) or the source (Figure 3.6b), and partially reflected and absorbed at the load.

It is very seldom practical to have a purely resistive load in the circuit as described above. More appropiately would be a reacative energy locus as described in the paragraph 3.3, where the same connection structure is capacitive and inductive during different parts of the switching cycle. One of the greatest challenges of electromagnetic design of a circuit would be to utilise such a locus to build "structural snubbers" which do not require capacitors or inductors.

3.7 A CONNECTION STRUCTURE AS A GENERAL PURPOSE POWER CONDITIONING ELEMENT.

Modelling of a connection structure as a energy flux channel identifies three mechanisms of its operation. The first and of primary importance is the Poynting vector, composed of a E and H field, which is responsible for the energy flow across the connection. The other two are secondary mechanisms consisting of the parasitic effects of ohmic losses in the conductor leads, and reactive power due to excessive magnetic or electric field intensities due to mismatched V/I ratios. An alternative to a circuit diagram would be an "energy diagram", compounded of series and parallel combinations of connection structures. In this context broader meaning is ascribed to a "connection structure", which is consequently is renamed as "power conditioning element", and is represented as a energy flow block as shown in Figure 3.7. Each block can be characterised by the following attributes:

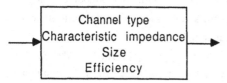

Figure 3.7 Defining a power conditioning element

a) CONNECTION TYPE:

   An element can either be an interconnection between two ore more other elements, or a single ended type. A single ended connection could be a reactive component, a dissipator or a source or sink of alternative energy forms, such as a battery or the mechanical energy conversion of an electric machine.

b) CHANNEL TYPE:

   The designation "channel" is due to the fact that energy flux is guided from a source to a destination by the circuit. Two channel types can be identified; the first being a conductor channel, which is a normal electric circuit connection. The second type is a magnetic channel, which is present in transformers and electric machines as described in the previous chapter.

c) CHARACTERISTIC IMPEDANCE:

   The characteristic impedance plays very much the same role as in transmission lines and microstrips, as it regulates the rate of energy flow and the amount of reactive energy, at different positions a in a circuit. Note that it is possible to allocate impedance values to chemical and mechanical sources and sinks of electrical energy. In many instances it is possible to model semiconductors as variable impedances.

d) SIZE:

The size is an indication of the energy capacity of a power conditioning block. Having size as a parameter is necessary, because circuit impedance is not equivalent to characteristic impedance. The characteristic impedance of a capacitor, for example, is a function of the dielectric and the width and distance between the plates, but not its length. The lenght of the parallel plates will be handled by the size parameter in this energy flux approach to circuit analysis.

e) EFFICIENCY:

Efficiency is the ratio of average output power to input power of an element, or , qualitatively speaking, a figure of merit of the irrecoverable losses. In a conductor channel it consists mainly of eddy current and conductions losses, whereas in a magnetic channel it would be chiefly hysteresis and eddy current losses.

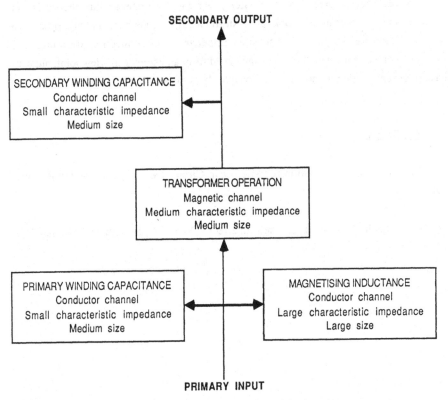

Figure 3.8 Modelling of a transformer with power conditioning elements

A structure such as a capacitor which consists of only two parallel plates, can, because of its constructional simplicity, be represented as one single ended power conditioning element. Figure 3.8 illustrates how one can typically model the more complex geometrical structures a transformer. The magnetising inductance part is presented as a single ended element, while the inter turn capacitance is modelled as a seperate element because it involves a different configuration and geometry than the inductor structure. The transformer operation takes place in the magnetic channel which is connected to the secondary.

As it has been stated at the start of the previous chapter on the Poynting vector; it is not envisaged the energy flux alternative to circuit modelling would cause a revolution in design procedures. Modelling of circuits in terms of power conditioning elements still needs considerable developement before it can be a practical alternative and will undoubtedly involve a more complex theory than conventional circuit analyses, and will have to rely a great extend on computer programs. The energy flux theory is not developed any further in this book, but instead calculation methods and algorithms are developed to analyse parasitic and stray impedances in connection structures and magnetic components, which can be appreciated from an energy flux viewpoint, but also have direct application in more conventional analysis of power converters.

3.8 REFERENCES

[1] T.C. Edwards; "Foundation for Microstrip Circuit Design", John Wiley & Sons, 1981.

[2] H.A. Wheeler; "Formulas for the skin effect", Proc. IRE, September 1942, pp. 412-424.

CHAPTER 4

EDDY CURRENTS IN CONDUCTORS

From a mathematical/physical point of view, the eddy current phenomena is governed by Maxwell's equations. It is shown that eddy currents can be described by differential equations appropriately called diffusion equations. Eddy currents in conductors are investigated, as they establish themselves in power electronic circuits. The pupose of this chapter is to derive the electromagnetic theory of eddy currents on which subsequent analyses are based.

4.1 THE EDDY CURRENT PHENOMENA

Electromagnetic field computation and modelling of eddy current related phenomena have received a lot of attention since alternating current has become the common mode of power supply for electrical apparatus. Eddy current is the collective term for the redistribution of alternating current in conductors as a function of frequency (skin effect), and the phenomena where one circuit carrying alternating current can induce circulating currents, without making ohmic contact, in any conductive material in the immediate vicinity of the circuit (proximity effect). Since the birth of electrical engineering, it has been a major topic in the design of power transformers and electric machines. It is perhaps the most important source of losses in these apparatus and has therefore direct bearing on its size, and the efficiency of power transmission and conversion, which is directly related to costs.

Minimising the eddy current losses involves the design optimisation of three dimensional geometric structures and magnetic materials with non-linear characteristics. Solving the field equations under these conditions is by no means an easy task. Before the age of fast electronic computers the approach to eddy current problems has been predominantly closed form analytical in nature. As large computers became available for engineering calculations, numerical methods such as finite differences emerged. Finite difference methods are not as widely used today as they have once been, as during the past fifteen years they made way to finite element methods, which have become the leading engineering tools. More recently boundary element methods and hybrid finite element methods appeared as promising alternative approaches.

Although many powerful finite element eddy current programs are available today, it has not really benefited the power converter circuit engineer. Programs of this nature are generally utilised for the design of magnetic converters (inductors, transformers and machines,) in the range of hundred kilowatts and upwards. The numerical nature of the solutions require large computers, large memory, significant computation speed and appreciable computation time, so that it is an expensive design tool. Speaking rather loosely from a mathematical point of view, an analytical method yields an algebraic solution in which the problem parameters appear as variables, the advantage being that the effect of altering one or more parameters is fairly readily appreciated. However, when mathematical models become too complicated to handle analytically, recourse is made to numerical methods, which have wider application but the disadvantage that the parameters are now concealed within the numerical results produced from a given set of data.

For the course of this study, the analyses are limited to eddy currents that occur in conductors, and magnetic cores are not considered. The reason being, firstly that magnetic materials such as ferrites and amorphous metals are on the market which have low losses at high frequencies, thus accentuating winding losses as a critical design factor. Secondly, the circuit engineer usually has very little say in the characteristics of the cores and have to rely on the manufacturers data, whereas the design of the windings is usually part of his cicuit design.

The purpose of this chapter is to outline the vector algebra framework for numerical and analytical eddy current analysis. From Maxwells equations the general forms of diffusion equations are derived, which are applicable to eddy current problems.Then finally, one dimension analytical equations for eddy currents, losses and impedances, of round and rectangular conductors are derived.

4.2 APPLICATION OF MAXWELL'S EQUATIONS

In this paragraph Maxwell's equations are manipulated into second order differential equations, which can be applied to conductors in circuits. The material under observation is assumed to be linear and homogeneous, and we do a frequency domain analysis by assuming sinusoidal waveforms and setting $\partial/\partial t = j\omega$. Maxwell's equations then become:

$$\nabla.E = \rho/\varepsilon \tag{4.1}$$

$$\nabla \times E = -j\omega B \tag{4.2}$$

$$\nabla.B = 0 \tag{4.3}$$

$$\nabla \times B = j\omega\varepsilon\mu E + \mu J \tag{4.4}$$

Next we rewrite above equations into second order differential format, in order to separate the electric and magnetic field quantities.

Set $J = \sigma E$ and substitute in equation (4.4):

$$\nabla \times B = (\sigma + j\omega\varepsilon)\mu E \tag{4.5}$$

Substitute (4.2) in (4.5):

$$\nabla \times \nabla \times E = \nabla(\nabla.E) - \nabla^2 E$$

$$= -(\sigma + j\omega\varepsilon)j\omega\mu E$$

Insert (4.1) in above equation and it becomes:

$$\nabla^2 E = (\nabla\rho)/\varepsilon + (\sigma + j\omega\varepsilon)j\omega\mu E \tag{4.6}$$

In the same manner by substituting (4.4) in (4.2) and rearranging, an expression for B can be obtained:

$$\nabla \times \nabla \times B = \nabla(\nabla.B) - \nabla^2 B$$

$$= -(\sigma + j\omega\varepsilon)j\omega\mu B$$

Since $\nabla. B = 0$

$$\nabla^2 B = (\sigma + j\omega\varepsilon)j\omega\mu B \tag{4.7}$$

Equations (4.6) and (4.7) are identical except for the term $(\nabla\rho)/\varepsilon$. In conductors this term can be associated with a charge distribution perpendicular to the current flow direction, which opposes external quasi-static electric fields. The interwinding electric field in coils, for example, will give rise to a $(\nabla\rho)/\varepsilon$ term. Common to equations (4.6) and (4.7) is the term $-\omega^2\varepsilon\mu$ which describes the effect of the displacement current. Displacement current is for all practical purposes negligible in the conductors themselves, but is essential in modelling capacitive currents between connection leads and between adjacent turns in coils. The remaining term, $j\omega\sigma\mu$ is associated with moving charge and which includes any eddy current effect.

Alternatively Maxwell's equations can be expressed in terms of potentials[13]. The magnetic vector potential, $\mathbf{A}$, and electric scalar potential, ϕ, is defined in accordance with (4.2) and (4.3) as follows:

$$\mathbf{E} = -\nabla\phi - j\omega\,\mathbf{A} \tag{4.8}$$

$$\mathbf{B} = \nabla \times \mathbf{A} \tag{4.9}$$

Substitute (4.8) in (4.1):

$$\nabla^2\phi + j\omega(\nabla.\mathbf{A}) = -\rho/\varepsilon \tag{4.10}$$

Substitute (4.8) and (4.9) in (4.1):

$$\nabla^2\mathbf{A} - \varepsilon\mu\omega^2\mathbf{A} - \nabla(\nabla.\mathbf{A} + j\omega\varepsilon\mu\phi) = -\mu\mathbf{J} \tag{4.11}$$

Equations (4.10) and (4.11) can be decoupled using the Lorentz condition:

$$\nabla.\mathbf{A} + j\omega\varepsilon\mu\phi = 0 \tag{4.12}$$

Equations (4.10) and (4.11) then becomes:

$$\nabla^2\phi - \varepsilon\mu\omega^2\phi = -\rho/\varepsilon \tag{4.13}$$

$$\nabla^2\mathbf{A} - \varepsilon\mu\omega^2\,\mathbf{A} = -\mu\,\mathbf{J} \tag{4.14}$$

Displacement current caused by time varying electric scalar potential is given by the term, $-j\omega\varepsilon\,\nabla\phi$. The Lorentz condition indicate that displacement current $j\omega\varepsilon\mu\,\nabla\phi$ causes a magnetic field and vector potential, such that the displacement current equals $\nabla(\nabla.\mathbf{A})$. Equation (4.13) is quite illuminating, as it identifies two types of charge, the one given by $-\varepsilon\nabla^2\phi$ and the other by $-\varepsilon^2\mu\omega^2\phi$. The first is a "static" charge, while the second is a "dynamic" charge which is observed because of a time varying scalar potential. Equation (4.14) is virtually identical to (4.13), with scalar quantities ϕ and ρ replaced by vector quantities $\mathbf{A}$ and $\mathbf{J}$. The term, $\omega^2\varepsilon\mathbf{A}$ can be related to "magnetic induced" displacement current, while $\nabla^2\mathbf{A}/\mu$ describe the applied current, combined with eddy currents.

4. 3 NUMERICAL METHODS[4-9]

Application of numerical methods to eddy current problems has been receiving a lot of attention in this current age of fast computers. A great variety of algorithms and methodologies are presented by authors in leading magazines such as the IEEE Transactions on Magnetics. In principle however, the solutions revolve around the equations derived in the previous paragraph.

Displacement currents are generally ignored in the calculation of eddy currents, making it a quasi-stationary effect, which is a realistic supposition provided the conductivity is high enough compared to the operating frequency. From a numerical method viewpoint, it is easier to work with potentials instead of fields, because the only requirement that a potential description must comply with is that the function must be single valued, (apart from the specific boundary conditions of the problem). A good example is the constraint that the divergence of a magnetic field vector has to be zero, while it follows automatically for the vector potential since $\nabla.\nabla \times \mathbf{A} = 0$ is a vector identity.

For computing eddy currents, it is necessary to work simultaneously with two potentials; usually it is the magnetic vector potential in conjunction with the electric scalar potential. Although the magnetic scalar potential lends itself in principle only to problems where current carrying conductors are absent, techniques do exist that enables one to solve eddy current problems with the magnetic scalar potential.

If one ignores displacement current, equation (4.14) simplifies to:

$$-\nabla^2\mathbf{A}/\mu = \mathbf{J} \qquad\qquad\qquad (4.15)$$

If (4.8) is substituted in $\mathbf{J} = \sigma\mathbf{E}$, the current density can be composed into two components, a source (or applied) component and an induced (or eddy current) component, that is:

$$\mathbf{J} = \mathbf{J_s} + \mathbf{J_i}$$

$$= -\sigma \nabla\phi - j\omega\sigma\mathbf{A} \qquad (4.16)$$

Substitute (4.16) in (4.15) and one obtains the vector potential equation that is being used in the majority of numerical algorithms to solve for eddy currents:

$$\nabla^2\mathbf{A}/\mu = \sigma \nabla\phi + j\omega\sigma\mathbf{A} \qquad (4.17)$$

The so-called T-Ω method has its origin in the defining equation

$$\mathbf{H} = \mathbf{T} - \nabla\Omega \qquad (4.18)$$

where $\mathbf{T}$ and Ω are the electric vector and magnetic scalar potentials, respectively. The curl or rotation of $\mathbf{H}$ equals that of $\mathbf{T}$, thus;

$$\nabla \times \mathbf{H} = \nabla \times \mathbf{T} = \mathbf{J} \qquad (4.19)$$

$\mathbf{T}$ can be regarded as a variation of the $\mathbf{H}$ vector since it has the same curl, but not the same divergence. $\mathbf{T}$ is the electric equivalent of the magnetic vector potential $\mathbf{A}$. Maxwells equations can be rewritten in terms of $\mathbf{T}$ and Ω, and a problem can be solved numerically by using for example the finite element method.

The thrust behind these numerical programs is to design 50/60 Hz magnetic components. At present, finite element programs for the solution of a range of 2-dimensional and quasi-3-dimensional (sometimes called "$2^1/_2$ - D" because of rotational symmetry and periodicity) field problems have reached a good degree of refinement, and are used extensively, both in the laboratory and by product companies. Substantial advances have been made in the development of techniques solving 3-dimensional problems, and some programs are already giving considerable help to design engineers[5].

4.4 ANALYTICAL METHODS

The tendency in analytical methods is to use field equations, as far as possible. This can be attributed to the boundary conditions being normally expressed in terms of magnetic and electric fields, and potential equations requiring an extra integration step. One also finds that field quantities may disappear in certain regions, while the derivative of a potential becomes zero while the function itself does not dissappear.

As for numerical methods, displacement currents are also ignored and (4.6) and (4.7) simplify to;

$$\nabla^2 E = j\omega\sigma\mu E \qquad\qquad\qquad\qquad (4.20)$$

$$\nabla^2 B = j\omega\sigma\mu B \qquad\qquad\qquad\qquad (4.21)$$

Set $J = \sigma E$ in (4.20):

$$\nabla^2 J = j\omega\sigma\mu J \qquad\qquad\qquad\qquad (4.22)$$

Analytical solutions are mostly based on diffusion equations (4.21) and (4.22), and in some cases (4.15). Solving the one dimensional problem is relatively straight forward and is done as described in the literature [1,2,10,11]. At first sight it may appear to be very artificial to examine the behaviour of eddy currents that are constrained to be functions of one coordinate direction only. However, in many instances the problem can be regarded as locally one dimensional, because the penetration distance is small compared with the other conductor dimensions. The classic problem of this type is the transformer and electrical machine lamination, which has a length and breadth much greater than its thickness. Another example of practical importance is the round conductor which become a one dimensional problem if solved in cylindrical coordinates.

A two dimensional problem is one in which the non-zero components of magnetic field and current density are dependent on one of the three coordinates used to define the system. Taking the rectangular coordinate system as a convenient example, we have the following possibilities:
  - the current flows in the z-direction only, and J_z, H_x and H_y are functions of x and y.
  - the magnetic field has a single component in the z-direction, and H_z, J_x and J_y are functions of x and y.

The first case is typically applicable to current carrying conductors, and equations (4.15) and (4.22) can be used to solve such a problem. In the second case the problem is usually solved the easiest by using (4.21).

One dimensional problems generally have closed form analytical solutions which are easy to manipulate and give good insight into the problem. Solving two-dimensional problems tend to be difficult and unless the boundary conditions are simple and symmetrical, it can easily become unsolvable. Due to its limited scope, analytical solutions have very much been neglected in recent times, while great progress has been made in numerical techniques.

4.5 ONE DIMENSIONAL SOLUTIONS

This paragraph describes some useful one dimensional analytical solutions of conductor leads which are utilised in subsequent chapters. For the purpose of this work we limit ourselves to nonmagnetic materials since circuit conductors are mostly made of materials such as copper, brass and aluminium, all of which are non-magnetic.

4.5.1 Flat current carrying strips[1]

The equations as they appear in reference [1] are rewritten in slightly different format in this and the next sub-paragraph.

Since $\mathbf{B} = \mu_0 \mathbf{H}$ inside the conductors we can rewrite equation (4.21) as;

$$\nabla^2 \mathbf{H} = \alpha^2 \mathbf{H} \tag{4.23}$$

where

$$\alpha = (1 + j)/\delta \tag{4.24}$$

and

$$\delta \equiv \text{skin depth}$$

$$= \frac{1}{\sqrt{\pi f \sigma \mu_0}} \qquad (4.25)$$

Equation (4.23) has a solution of the form

$$H_z = K_1 e^{\alpha y} + K_2 e^{-\alpha y} \qquad (4.26)$$

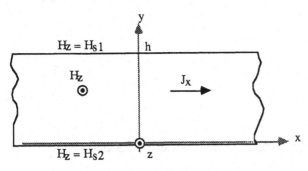

Fig. 4.1 Cross section of a semi-infinite plate

Let the strip have a height, h, and width, w, and the coordinate system be as indicated in Fig. 3.1. In case of the plate carrying a current I, application of Ampere's circuital law yields the following boundry values:

$$H_{s1} = -H_{s2} = \frac{I}{w} \qquad (4.27)$$

Solving for H yields:

$$H_z = \frac{I \cosh \alpha y}{w \sinh \alpha h/2} \qquad (4.28)$$

Since $dH_z/dy = J_x$ we get:

$$J_x = \frac{\alpha I \cosh \alpha y}{2w \sinh \alpha h/2} \qquad (4.29)$$

The loss per unit length of conductor is given by:

$$P_{1s} = \frac{w}{2\sigma} \int_0^h |J_x^2| \, dy$$

$$= \frac{I^2}{4w\sigma\delta} \frac{\sinh \upsilon + \sin \upsilon}{\cosh \upsilon - \cos \upsilon} \tag{4.30}$$

$$\text{with } \upsilon = \frac{h}{\delta} \tag{4.31}$$

4.5.2 Losses in a strip conductor subjected to a uniform magnetic field[1]

The applied field H_s is the same on both surfaces and (4.26) becomes:

$$H_z = H_s \frac{\cosh \alpha y}{\cosh \alpha h/2} \tag{4.32}$$

The current density is:

$$J_x = \alpha H_s \frac{\sinh \alpha y}{\cosh \alpha h/2} \tag{4.33}$$

The loss unit per length is given by:

$$P_{1p} = \frac{w}{2\sigma} \int_{0}^{h} |J_x^2| \, dy$$

$$= \frac{w \, H_s^2}{\sigma \, \delta} \frac{\sinh \upsilon - \sin \upsilon}{\cosh \upsilon + \cos \upsilon} \tag{4.34}$$

4.5.3 Current distribution and impedance of a strip conductor with arbitrary field at the surfaces

More general equations,based on the method followed by Dowell[11], are now derived. We do not calculate losses but instead derive the impedance of the strip conductor. Note also that the method differs, instead of calculating the integrals of J^2 and H^2 we make use of the fact that no transversal electric field can exist inside the conductor. No limitation is placed on the values that H may assume on the two surfaces of the strip, provided they have the same phase. (This is in most cases true with the exception of a few specialised structures such as the coaxial sjunt.)

Diffusion equation (4.22) can be simplified to:

$$d^2J_x/dx^2 = \alpha^2 J_x \tag{4.35}$$

where α is given by (4.24)

The general solution of (4.35) is:

$$J_x = P \cosh \alpha y + Q \sinh \alpha y \tag{4.36}$$

Application of Ampere's yield the following boundary values;

$$H_{s2} - H_{s1} = \frac{I}{w} \tag{4.37}$$

Let $H_{s1} = \dfrac{(k-1)I}{w}$ (4.38)

$H_{s2} = \dfrac{kI}{w}$ (4.39)

where k is any real number.

The magnetic field intensity inside the strip conductor is therefore given by:

$$H_z(y) = \frac{1}{w} \left[I(k-1) + \int_0^y w\,J_x\,dy \right]$$ (4.40)

The magnetic flux density is:

$$B_z(y) = \mu_0 \left[\frac{I(k-1)}{w} + \int_0^y J_x\,dy \right]$$ (4.41)

The electric field along the length is composed of an inductive and resistive voltage drop; i.e.

$$E_x = \rho J_x + j\omega\,(\Phi_\iota + \Phi_e)$$ (4.42)

where Φ_ι and Φ_e are the internal and external flux linkage respectively. Equation (4.42) can be simplified by differentiating it with respect to y and utilising the fact that E_x is constant over the height of the conductors.

$$0 = dE_x/dy = \rho\,dJ_x/dy + j\omega\,d(\Phi_\iota + \Phi_e)/dy$$ (4.43)

since Φ_e is constant and $d\Phi_\iota / dy = B_z$, we get:

$$dJ_x/dy = -j\omega\sigma B_z$$ (4.44)

Substitute (4.44) in (4.41):

$$dJ_x/dy = j\omega\mu_0\sigma \left[\frac{I(k-1)}{w} + \int_0^y J_x \, dy \right] \tag{4.45}$$

Set $y = 0$ in (4.45) and we get:

$$Q = \frac{\alpha I(k-1)}{w} \tag{4.46}$$

The total current is simply the integral of the current density;

$$I = w \int_0^h J_x \, dx \tag{4.47}$$

and we obtain:

$$P = \frac{\alpha I}{w} \left[\frac{1}{\sinh \alpha h} + (1-k)\tanh \frac{\alpha h}{2} \right] \tag{4.48}$$

Equation (4.36) then becomes:

$$J_x = \frac{\alpha I}{w} \left[\frac{\cosh \alpha x}{\sinh \alpha h} - (k-1) \tan \frac{\alpha h}{2} + (k-1) \sinh \alpha x \right] \tag{4.49}$$

Hence at the top (y=h) the current density is given by:

$$J_{x\,top} = \frac{\alpha I}{w} \left[\coth \alpha h + (k-1) \tanh \frac{\alpha h}{2} \right] \tag{4.50}$$

The voltage drop per unit length on the surface of the conductor is;

$$E_x = \frac{J_{x \, top}}{\sigma} \tag{4.51}$$

Divide both sides of equation (4.51) with I, and we get the internal impedance per unit length:

$$Z_i = \frac{1}{\sigma w \delta} \left[\left\{ \frac{\sinh 2\upsilon + \sin 2\upsilon}{\cosh 2\upsilon - \cos 2\upsilon} + (k-1) \frac{\sinh \upsilon - \sin \upsilon}{\cosh \upsilon + \cos \upsilon} \right\} \right.$$

$$\left. + \; j \left\{ \frac{\sinh 2\upsilon - \sin 2\upsilon}{\cosh 2\upsilon - \cos 2\upsilon} + (k-1) \frac{\sinh \upsilon + \sin \upsilon}{\cosh \upsilon + \cos \upsilon} \right\} \right] \tag{4.52}$$

4.5.4 Impedance of a cylindrical conductor[2]

Let us consider a solid cylinder carrying alternating current in the axial direction. If the length of this conductor with respect to its cross section is large, as it is usual with wires, we do not have to consider the effects of its end on the current distribution. Assuming the return path to be infinitely far removed we have a case of cylindrical symmetry with the magnetic field having only a tangential component and the current density only an axial one. The subsequent equations give a shortened version of the detailed derivation as found in Reference [2].

Writing equation (4.22) in cylindrical coordinates;

$$d^2J_z/dr^2 + 1/r \; dJ_z/dr = j\omega\sigma\mu J_z \tag{4.53}$$

This is a Bessel differential equation. The unique solution over the cross section of the conductor will be:

$$J_z = C \, J_0(\alpha r) \tag{4.54}$$

The integration constant, C, is solved by setting the integral over the cross section equal to the total current, giving;

$$J_z = \frac{\alpha I\, \mathbb{J}_0(\alpha r)}{2\pi r_0\, \mathbb{J}_1(\alpha r_0)} \tag{4.55}$$

where $r_0 \equiv$ the radius of the cylinder.

In a similar manner to par. 4.5.3, we take the voltage drop on the surface $(r = r_0)$;

$$E_z = \frac{J_{z\ \text{surface}}}{\sigma} \tag{4.56}$$

and by dividing by I, the internal impedance becomes;

$$Z_i = \left(\frac{\text{ber}(\zeta r_0)\text{bei}'(\zeta r_0) - \text{bei}(\zeta r_0)\text{ber}'(\zeta r_0)) + j(\text{ber}(\zeta r_0)\text{ber}'(\zeta r_0) + \text{bei}(\zeta r_0)\text{bei}'(\zeta r_0)}{\sqrt{2}\ \pi r_0 \sigma \delta \quad \{\text{ber}'\,^2(\zeta r_0) + \text{bei}'\,^2(\zeta r_0)\}} \right)$$

where $\zeta = \dfrac{\sqrt{2}}{\delta}$ \hfill (4.57)

4.5.5 Losses in a cylindrical conductor due to a transverse magnetic field

A cylinder of radius r_0, conductivity σ is placed in a homogeneous magnetic field H_0. The solution involves much more theory than the previous cases. The complete derivation is covered in reference [2]. Note that an error was made in the last stage of the derivation when the Lommel integral was taken (equations 7-51, 52 on page 96).

Equation (4.17) is convenient to use since the source current component is zero. It simplifies to the following vector potential diffusion equation:

$$\partial A_z/\partial r^2 + 1/r\ \partial A_z/\partial r + 1/r^2\ \partial^2 A_z/\partial \theta^2 = \alpha^2 A_z \tag{4.58}$$

Through quite lengthy vector algebra the expression for the vector potential inside the cylinder is found to be:

$$A_z = \frac{4 \mu_0{}^2 H_0 \; \mathbb{J}_1(j^{3/2}\zeta r) \sin \theta}{j^{3/2} \; \zeta \; \mathbb{F}(j^{3/2}\zeta r_0}$$

(4.59)

The current density in terms of the vector potential is:

$$J_z = \sigma E_z = -j\omega\sigma A_z$$

$$= 4 \mu_0 H_0 j^{3/2}\zeta \; \mathbb{J}_1(j^{3/2}\zeta r)\sin \theta / \mathbb{F}(j^{3/2}\zeta r_0)$$

(4.60)

The eddy current losses per unit length of the cylinder is given by the double integral:

$$P = \frac{1}{2\sigma} \int_0^{r_0} \int_0^{2\pi} |J_z|^2 \, r \, dr \, d\theta$$

$$= \frac{-2\pi \, \zeta r_0 H_0{}^2}{\sigma} \; \frac{ber_2(\zeta r_0) \; ber'(\zeta r_0) + bei'(\zeta r_0) \; bei_2(\zeta r_0)}{\{ber^2(\zeta r_0) + bei^2(\zeta r_0)\}}$$

(4.61)

4.6 REFERENCES

[1] RL Stoll; "The analysis of eddy currents."; Clarenden Press - Oxford; 1974.

[2] J Lammeraner, M Stafl; "Eddy Currents"; Iliffe Books - London; 1966.

[3] T Tortschanoff; "Survey of numerical methods in field equation."; IEE Trans. on Magnetics, Vol Mag-20, (5), Sept. 1984, pp 1912 - 1917.

[4] A Konrad; "Eddy currents and modelling."; IEEE Trans. on Magnetics; Vol Mag-21, (5), Sept 1985, pp1805-1810.

[5] RJ Lari, LR Turner; "Survey of eddy current programs."; IEEE Trans. on Magnetics; Vol Mag-19, (6), Nov 1983, pp2474-2477.

[6] TW Preston, ABJ Reece; "The contribution of the finite-element method for the design of electrical machines."; IEEE Trans. on Magnetics, Vol Mag-19, (6), Nov. 1983, pp 2375 - 2380.

[7] M Riaz; "The circuit-field connection in two dimensional ac finite-element analysis of multiply-excited magnetic structures."; IEEE Trans. on Power Apparatus and Systems, Vol PAS-104, 970, July 1985. pp 1797 - 1803.

[8] MS Hwang, WM Grady; "Assesment of winding losses in transformers due to harmonic currents."; International Conference on Harmonics in Power Systems, Worcester Polytechnic Institute, Oct. 1984, pp 81 - 88.

[9] K Preis, H Stoegner, KR Richter; "Calculation of eddy current losses in air coils by finite element method."; IEEE Trans, on Magnetics, Vol Mag-18, (6), Nov. 1982, pp 1064 - 1066.

[10] HJ Kaul; "Stray-current losses in stranded windings of transformers."; Trans. of AIEE, Vol 6, (30), June 1957, pp 137 - 149.

[11] Pl Dowell; "Effects of eddy currents in transformer windings."; Proc. IEE, Vol 113, (8), Aug. 1966, pp 1387 - 1394.

[12] SA Swann, JW Salmon; "Effective resistance and reactance of rectangular conductor in a semi-closed slot."; Proc. IEE, Vol 110, (9); Sept. 1963.

[13] A Shadowitz; "The Electromagnetic Field"; McGraw-Hill, Kogusha, Ltd, 1975.

CHAPTER 5

STRUCTURAL IMPEDANCE OF POWER CONNECTIONS; ANALYSIS AND METHOD OF CALCULATION

Increasing the switching speed and power levels in converters brings one to a point where the parasitic impedance of interconnections begins to influence the efficiency and waveforms of the system. With this in mind, modelling of structural impedances of connections is investigated, and an algorithm is presented which calculates the values of stray inductance, resistance and capacitance for most practical connections encountered in power converters.

5.1 CONSIDERATIONS ON CALCULATION OF PARASITIC IMPEDANCES

Equations, graphs or techniques are generally not available to assist an engineer, faced with a stray impedance problem, to calculate its values or apply it to his circuit design. Related fields on this topic are high frequency signal transmission lines, low frequency power transmission lines and microstrip theory. Although some aspects of each of the above theories can be useful, not one is entirely suitable for this purpose. A difficulty common to all three is that their application is normally confined to simple sinusoidal waveforms, while one generally encounters more complex waveforms in power converters. In transmission lines and microstrips the frequencies are so high that the current flows only on the surface of the conductors, whereas this is not the case with power converter interconnections. Low frequency power transmission line theory[6] is also not very useful, since its geometrical scope is not sufficient for power connections and the low power model tends to neglect the capacitance between the lines.

Two parasitic impedance components can be identified within a connection structure, namely external and internal impedance. The external component is determined by the electric and magnetic fields around the outgoing and returning conductors, giving rise to the external capacitance and inductance. The internal impedance is governed by what happens inside the conductors. The current distribution inside the conductors gives rise to an internal magnetic field as well as ohmic losses, which in circuit terminology would be the internal inductance and resistance. One finds, for example, that the current distribution causes the resistance to increase and the internal inductance to decrease, the higher the frequency and the smaller the distance between the outgoing and returning current paths becomes (see Fig. 5.1 and description in par 1.2).

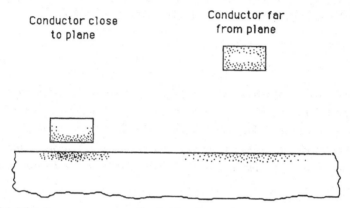

Conductor close
to plane

Conductor far
from plane

Figure 5.1 High frequency current distribution in conductors.

"Internal capacitance" seems to be a term which is not mentioned anywhere in literature, which indicates that it is very difficult to analyse or completely negligible or both of these. An internal capacitance would involve the dynamic distribution of charge inside a conductor, which contrary to the main current, one would intuitively expect to be a surface current, which is limited to the surfaces representing the shortest distance between the forward and return conductors. One would therefore expect it to be very small and for the purpose of our analysis we neglect the internal capacitance.

Figure 5.2 shows possible two dimensional cross sections of interconnections, composed of round, rectangular conductors and conductive planes. Two types of dielectric configurations are to be covered, namely where a flat dielectric slab is placed between the outgoing and return conductors, and where the conductors are completely submerged in the dielectric material (Fig. 5.3). The analysis of this chapter attempts to solve the parasitic impedance of these configurations, which should be sufficient to cover most interconnections in power converters. The analysis will be two dimensional, based on the cross section dimensions, while provision is also made to include end effects.

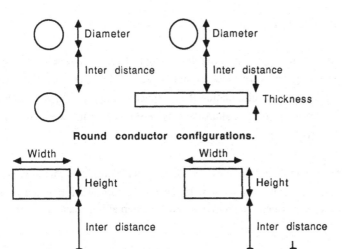

Round conductor configurations.

Rectangular conductor configurations.

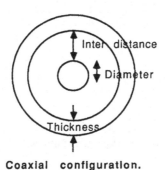

Coaxial configuration.

Figure 5.2 Conductor configurations.

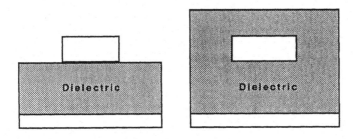

Slab   **Homogeneous**

Figure 5.3 Dielectric configurations.

5.2 EXTERNAL IMPEDANCE

The external impedance is solved by assuming the conductive material to be of infinite conductivity, thereby setting the internal electric field to zero. Current can only flow on the surface of the conductors, thus eliminating the presence of internal magnetic field, even at dc. The calculation procedure is almost exclusively based on the work done by H.A. Wheeler[4] for microstrip applications.

The external inductance and capacitance can be easily calculated if the characteristic impedance of the interconnection is known. If the conductors are immersed in a homogeneous dielectric, Z_0 is calculated for $\varepsilon_r=1$, and the external impedances are given by:

$$L_{ex} = \frac{Z_0}{c} \tag{5.1}$$

$$C = \frac{\varepsilon_r}{Z_0 c} \tag{5.2}$$

assuming $\mu_r = 1$.

The external impedances of an interconnection with a dielectric slab (Fig 5.3) can be calculated, using equations (5.6 to 12) for Z_0, as follows:

$$L_{ex} = \frac{Z_0}{v_p} \tag{5.3}$$

$$C = \frac{1}{Z_0 v_p} \tag{5.4}$$

where $v_p = \dfrac{c}{\sqrt{\varepsilon_{eff}}}$

An exact formula for the round wire without a dielectric slab is known[4], but instead we replace round conductors with square ones for the sake of standardising on a general procedure, and because the round wire formula is not valid when a dielectric slab is present. Setting the area of square conductors equal to a round conductor, we obtain:

$$w = t = \frac{\sqrt{\pi}\ d}{2} \qquad (5.5)$$

The computation of the characteristic impedance is an approximate method to meet design requirements within practical tolerance, based on some vigorous derivations by conformal mapping[4]. The physical width w is first corrected by amount δw in order to obtain the equivalent width w' which is to be used in an equation valid for thin strips.

$$Z_0 = \frac{42.4}{\sqrt{(\varepsilon_r+1)}}\ \ln\{1 + \frac{4h}{w'}\ [b + \sqrt{(b^2 + a\pi^2)}]\} \qquad (5.6)$$

$$w' = w + a\delta w \qquad (5.7)$$

$$b = \frac{(14 + 8/\varepsilon_r)\ (4h)}{11}\ \frac{}{w'} \qquad (5.8)$$

$$a = \frac{1 + 1/\varepsilon_r}{2} \qquad (5.9)$$

$$\delta w = \frac{t}{\pi}\ \{1 + \ln\ (\frac{4e}{\sqrt{[(t/h)^2 + (\pi\ (wt+1.1))^{-2}]}})\ \} \qquad (5.10)$$

If one end of the line is open circuit or terminated in a high impedance, the fringing effect of the field at the end will cause the line to appear to be longer than its physical length. The extra length can be calculated as follows[2]:

$$\delta l = 0.412h\ \frac{(\varepsilon_{eff} + 0.3\ )}{\varepsilon_{eff} - 0.258}\ \frac{(w/h + 0.262)}{w/h + 0.813} \qquad (5.11)$$

$$\text{where } \varepsilon_{eff} = \frac{1}{2} \left[1 + \frac{1}{\sqrt{(1 + 10^{h/w})}} \right] \qquad\qquad (5.12)$$

5.3 INTERNAL IMPEDANCE

Since the beginning of this century regular contributions on the solution of eddy current appeared in scientific and engineering journals. However, the mathematical difficulties are such that closed form analytic solutions which can solve the internal impedance of the configurations in Figure 5.2, (with the exception of the coaxial case) are not available today.

The internal impedance of a round conductor, very far removed from the current return path can be readily calculated (equation 4.57). If the proximity effect also plays a role in the current distribution, the solution becomes more complicated, as indicated by Carson[5]. Unfortunately also, his exhaustive derivation did not include the internal inductance, although he indicated that it can be done within the scope of his derivation. When it comes to solving current distribution inside rectangular conductors, the solutions become even more difficult. The solution for a rectangular conductor infinitely far removed from the current return path, as derived by Lammeraner and Stafl[1], is not only complicated, but involves summations over infinite mathematical series which seem converge slowly. The author was not able to find analytical solutions of the proximity effect in rectangular conductors. Another aspect which also has to be solved is that of the current distribution inside a conductive plane as governed by the proximity effect. To do this in closed form might be the most difficult task.

Rather than to turn to a full fledged numerical solution, the approach was to use various "simple" analytical expressions, which do not consume too much computation time, and to combine them into a scheme which gives a good approximation of the internal impedance. Three procedures can be distinguished within this algorithm:

    a) A general solution for skin effect impedance of round and rectangular conductors.

    b) A proximity effect adjustment factor.

    c) Solution of the internal impedance of the conductive plane.

5.3.1. Skin effect impedance of round and rectangular conductors

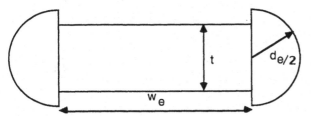

Figure 5.4 Composition of a rectangular conductor.

The internal impedance of a round conductor is obtained for equation (4.57). The rectangular conductor is composed of an equivalent assembly consisting of the two halves of a round conductor placed on either side of a one dimensional strip (see Fig 5.4). The internal impedance of the strip with dimensions w_e and t is evaluated by setting k = 1/2, in equation (4.52). At low frequencies the cross section area must be the same, whereby d_e and w_e can be expressed as follows:

$$d_e = \frac{2t}{\sqrt{\pi}} \qquad (5.13)$$

$$w_e = w - t \qquad (5.14)$$

At high frequencies when the penetration depth is much smaller than the conductor dimensions, the requirement is that the circumference of the cross section must be equal. Retaining condition (5.14), d_e becomes:

$$d_e = \frac{4t}{\pi} \qquad (5.15)$$

5.3.2. Proximity effect multiplication factor

Factors C_R and C_L are calculated, by which the skin effect resistance and inductance are respectively multiplied in order to allow for the influence of the proximity effect. These multiplication factors are obtained from the ratio of impedances at k = 1/2 and k = k' from equation (4.52), as follows:

$$C_R = \frac{R_i(k=k')}{R_i(k={}^1/_2)} \qquad (5.16)$$

$$C_L = \frac{L_i(k=k')}{L_i(k={}^1/_2)} \qquad (5.17)$$

The solution of k' is done, using an approximative method, involving an iterative routine. The influence of the current redistribution is incorporated by utilising uniform current distributions as indicated in Figure 5.5. The thickness of the current areas is δ, if δ is smaller than a/2, or a/2 if it is bigger.

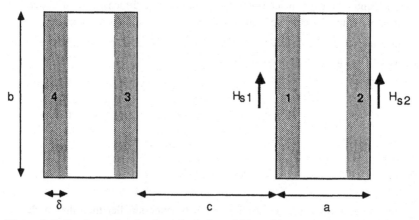

Figure 5.5 Configuration for the calculation of factor k'.

From equations (4.37-39) the total current is:

$$I = \frac{(H_{s2} - H_{s1})}{w} \qquad (5.18)$$

and

$$k' = \frac{1}{(1 + H_{s1}/H_{s2})} \qquad (5.19)$$

We assume the value of current in areas 1 and 4 in figure 5.5 to be;

$$I_1 = -I_4 = \frac{g}{(g + h)} \tag{5.20}$$

and in areas 2 and 3;

$$I_2 = -I_3 = \frac{h}{(g + h)} \tag{5.21}$$

where

$$g = (1-k') + k\,e^{-a/\delta} \tag{5.22}$$

$$h = k + (1-k')e^{-a/\delta} \tag{5.23}$$

Equations (5.18 to 23) in conjunction with (6.47 to 48) can be combined in an iterative procedure to calculate the value of k'. Note this method involves an approximation since the phase delay in the current density is not included in equation (5.22,23). This should however, not cause large errors.

5.3.3 Internal impedance of the conductive plane.

The closer a round or rectangular conductor is to the conductive plane, the more the return current flowing in the plane tends to concentrate itself directly underneath the conductor (see Figure 5.1). A (non-numerical) solution to this problem is the so-called "incremental-inductance" method originally proposed by H.A. Wheeler[3].

The incremental-inductance is a formula which gives the resistance caused by the skin effect, but is based entirely on inductance computations. Its great value lies in its general validity for all metal objects in which the current and magnetic intensity are governed by the skin effect. For a conductor with thickness at least a few times larger than the depth of penetration, the surface resistivity, R_1, defined as the resistance of a surface of equal length and width[3], as follows:

$$R_1 = \frac{1}{\delta\sigma}$$

(5.24)

The internal impedance of a conductor of circumference q and length l is given by;

$$Z_i = R_i + jX_i$$

with

$$R_i = X_i = R_1 \frac{l}{q}$$

(5.25)

Through some manipulation the internal inductance can be expressed as;

$$L_i = \frac{X_i}{\omega} = \frac{l \mu \delta}{2q}$$

(5.26)

which implies that the internal inductance of a conductor where the skin effect predominates, can be calculated from the difference in external inductance between a full size conductor and one where the outside surface is shrunk by $\delta/2$.

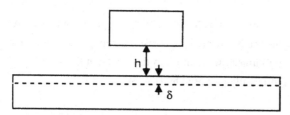

Figure 5.6 Incremental inductance rule on a plane conductor.

By varying the distance between the conductor and the plane by $\delta/2$ (Figure 5.6) the internal inductance is given by:

$$L_i = \frac{(Z_0 (h + \delta/2) - Z_0(h))}{c}$$

(5.27)

From (5.39), then:

$$R_i = \omega L_i \tag{5.28}$$

The incremental inductance method is unfortunately limited to high frequencies where the depth of penetration is significantly smaller than the conductive plane thickness, s. The low frequency internal impedance is solved by the following approximate method, when the depth of penetration is larger than 0.25s:

a) Calculate internal impedance for a penetration depth of 0.25s, denoted by $R_i^{\#}$ (0.25s) and $L_i^{\#}$ (0.25s) using (5.27) and (5.28).

b) Using equation (4.52) for a per unit width conductive strip, setting k = 1, for penetration depths of 0.25s and δ, denoting the impedances $R_i@(0.25s)$, $R_i@(\delta)$, $L_i @(0.25s)$ and $L_i @(\delta)$. From this we define multiplication factor K_R, K_L as follows:

$$K_R = \frac{R_i@(\delta)}{R_i@(0.25s)} \tag{5.29}$$

$$K_L = \frac{L_i@(\delta)}{L_i@(0.25s)} \tag{5.30}$$

c) The low frequency impedance is calculated from:

$$R_i(\delta) = K_R R_i^{\#}(0.25s) \tag{5.31}$$

$$L_i(\delta) = K_L L_i^{\#}(0.25s) \tag{5.32}$$

In this way we let the current distribution, along the width of the plane, at low frequencies be the same as at the frequency at which the penetration depth is a quarter of the plane thickness. Yet, the current distribution across the height of the plane, is the exact one dimensional solution given by equation (4.52).

5.4 PREDICTIONS AND EXPERIMENTAL MEASUREMENTS

In his 1921 paper[7] S Butterworth refers to an extensive series of measurements on the resistance of a go-and-return system of parallel conductors, by Kennelly, Laws and Pierce[8] The wire used was copper of diameter 1.168cm, and the frequency employed ranged from 60Hz to 5kHz. The spacing was varied from 3 to 600mm, so that the observations provide a fairly complete check for the twin wire part of the algorithm described in this chapter. The next five graphs display the ac to dc resistance ratio of the predictions compared to the measured values.

From the graphs it can be seen that very good agreement is obtained for large spacing values, but it deteriorates at high frequencies and small spacings. Some deviation is to be expected, especially for the round conductor case, since the algorithm uses a hybrid method where the proximity effect is calculated for an "equivalent" square shaped conductor. Despite the errors observed in the above graphs, the predictions by the algorithm should still be useful, and for rectangular conductors, particularly for strip conductors, better ac resistance predictions can be expected.

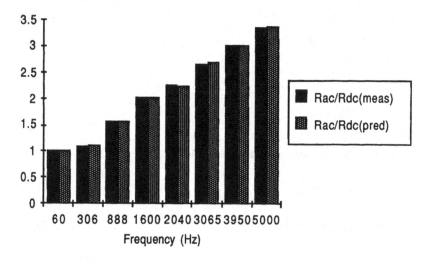

Figure 5.7 Resistance ratio for spacing of 600mm.

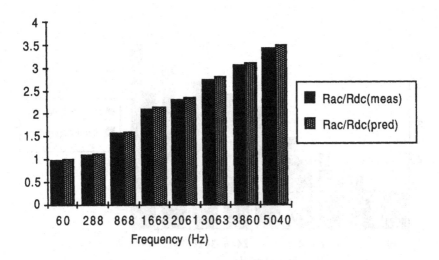

Figure 5.8 Resistance ratio or spacing of 200mm.

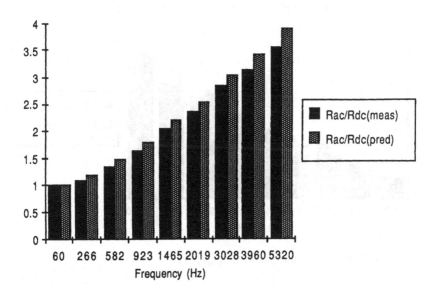

Figure 5.9 Resistance ratio for spacing of 64mm.

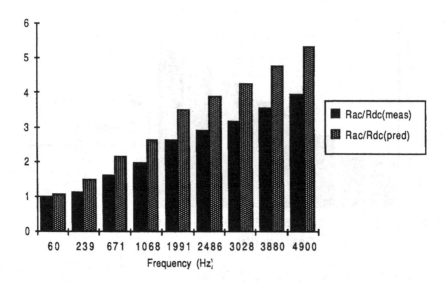

Figure 5.10 Resistance ratio for spacing of 8mm.

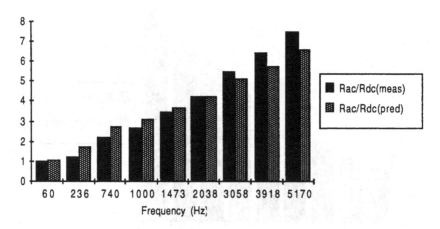

Figure 5.11 Resistance ratio for spacing of 3mm.

5.5 REFERENCES

[1] J Lammeraner, M. Stafl; "Eddy Currents", Iliffe Books-London, 1966.

[2] TC Edwards; "Foundation for Microstrip Circuit Design", John Wiley & Sons, 1981.

[3] HA Wheeler; "Formulas for the skin effect", Proc. IRE, September 1942, pp. 412-424.

[4] HA Wheeler; "Transmission-line Properties of a strip on a Dielectric Sheet on a Plane", IEEE Trans. on Microwave Theory and Techniques.; August 1977, pp. 631-647.

[5] JR Carson; "Wave Propogation over Parallel Wires: the Proximity Effect", Philosophical Magazine, vol. 41, April 1921, pp. 607-632.

[6] OI Elgerd; "Electric Energy Systems Theory: An Introduction": McGraw Hill 1971.

[7] S Butterworth; "Eddy-Current Losses in Cylindrical Conductors, with Special Applications to the Alternating Current Resistance of Short Coils"; Philosophical Transactions, Vol. 222, 1921, pp 57-100.

[8] Kennelly, Laws and Pierce; Trans. American Inst. El. Eng., Volume 35, Part 2, 1915.

CHAPTER 6

SKIN AND PROXIMITY EFFECT LOSSES IN TRANSFORMER AND INDUCTOR WINDINGS

Equations are derived to calculate conductive and eddy current losses in transformer and inductor windings. A prominent feature of this mathematical approach is that it permits seperation of the ohmic losses into skin effect, proximity effect and frequency components.

6.1 ORTHOGONALITY OF CURRENT DENSITY DISTRIBUTION

The approach being followed in this chapter, is essentially a further developement of the technique employed by Lammeraner in reference [1], where he derives an expression for the high frequency resistance of a single isolated litz wire. One factor, which is the outstanding feature of the mathematical derivation of this chapter, is occurrences of orthogonality in the power integrals which decouples all the essential dissipation sources. It will be shown that it is possible to determine the total winding losses for an arbitrary periodic current waveform by calculating the skin effect and proximity effect losses seperately for each frequency component and simply adding all the terms together.

The current flow inside a conductor lead is taken to be in the z-direction and the density distribution is only a function of the x,y coordinates, thus simplifying the problem to a two dimensional one. If the current density is expanded into a Fourier series we get;

$$J(x,y,t) = J_1(x,y) \cos \omega t + J_2(x,y) \cos 2\omega t + ... + J_i(x,y) \cos i\omega t + ... \quad (6.1)$$

where J_i = i'th Fourier coefficient of current density (peak value phasor quantity),

   ω = repetition frequency of the current waveform.

The power dissipation per unit length of the conductor is by definition:

$$P_1 = \frac{1}{T\sigma} \int_A \int_0^T |\, J(x,y,t)|^2 \, dt \; dA \tag{6.2}$$

Since the time integral of the product of unequal Fourier components over a period equals zero, (6.2) can be simplified to:

$$P_1 = \frac{1}{2\sigma} \sum_{i=1}^{\infty} \int_A J_i \, J_i^* \, dA \tag{6.3}$$

Eddy current in conductive leads can be subdivided into two components, namely a skin effect component and a proximity effect component. The first component entails the redistribution of the current density towards the surface of a conductor, infinitely far removed from the current return path, which carries alternating current. The proximity effect is associated with circulating eddy current components induced in conductive material by magnetic fields from nearby current carrying conductors. Note that the integral of proximity effect current density over the cross section area of the conductor is by definition always zero.

Designating the skin effect and proximity effect current densities J_s and J_p respectively, equation (6.3) becomes:

$$P_1 = \frac{1}{2\sigma} \sum_{i=1}^{\infty} \int_A (J_{si} + J_{pi})(J_{si}^* + J_{pi}^*) \, dA \tag{6.4}$$

The above equation can be simplified if we limit the scope of the eddy current analysis to the one dimensional, symmetric shapes covered in paragraph 4.5, namely, cylindrical wire and strip conductor, which have an axis and a plane of symmetry respectively. In this case J_p displays a reciprocal symmetry to that of J_s. J_s is an even function, while J_p is odd if the external magnetic field is perpendicular to the symmetry plane/axis. Under these conditions equation (6.4) becomes:

$$P_1 = \frac{1}{2\sigma} \sum_{i=1}^{\infty} \int_A (J_{si} J_{si}{}^* + J_{pi} J_{pi}{}^*) \, dA \qquad (6.5)$$

$$= \sum_{i=1}^{\infty} (P_{1si} + P_{1pi}) \qquad (6.6)$$

where P_{1si} = skin effect losses at the i'th harmonic

$$= \frac{1}{2\sigma} \int_A J_{si} J_{si}{}^* \, dA \qquad (6.7)$$

P_{1pi} = proximity effect losses at the i'th harmonic

$$= \frac{1}{2\sigma} \int_A J_{pi} J_{pi}{}^* \, dA \qquad (6.8)$$

6.2 EDDY LOSSES IN ROUND AND STRIP CONDUCTORS

For round and strip conductors skin effect losses per unit length can be expressed (see equations (4.30) and (4.57)) as;

$$P_{1s} = F \, I^2 \qquad (6.9)$$

and proximity effect losses per unit lenght, as from equations (4.34) and (4.61);

$$P_{1p} = G \, H_e^2 \qquad (6.10)$$

where   I   = the current in the conductor

     H_e = the external magnetic field caused by surrounding currents

In the case of strip conductors, only the component of magnetic field parallel to the surface is applicable. For the purpose of calculating coefficients F and G in computer programs (, see Appendixes C, D), they can be expressed in terms of infinite series as follows:

$$F_{strip} = \frac{R_{dc}\,\gamma}{4} \frac{(\sinh\gamma + \sin\gamma)}{(\cosh\gamma - \cos\gamma)}$$

$$= \frac{R_{dc}}{4} \sum_{n=1}^{\infty} \left[\frac{\gamma^{4n-1}}{(4n-1)!} \right] \Bigg/ \sum_{n=1}^{\infty} \left[\frac{\gamma^{4n-2}}{(4n-2)!} \right] \qquad (6.11)$$

$$G_{strip} = \frac{w\gamma}{h\sigma} \frac{(\sinh\gamma - \sin\gamma)}{(\cosh\gamma + \cos\gamma)}$$

$$= \frac{w\gamma}{h\sigma} \sum_{n=1}^{\infty} \left[\frac{\gamma^{4n-1}}{(4n-1)!} \right] \Bigg/ \sum_{n=1}^{\infty} \left[1 + \frac{\gamma^{4n}}{4n!} \right] \qquad (6.12)$$

$$\text{where} \quad \gamma = \frac{h}{\delta} \qquad (6.13)$$

$$R_{dc} = \frac{1}{hw\sigma} \qquad (6.14)$$

with h = height of strip
      w = width of strip

The expressions applicable to round conductors are as follows:

$$F_{round} = \frac{R_{dc}\gamma}{4} \frac{(ber(\gamma)bei'(\gamma) - bei\,(\gamma)ber'(\gamma))}{ber'^2(\gamma) + bei'^2(\gamma)}$$

$$= \frac{R_{dc}\gamma}{4} \left[-\frac{\gamma}{2} \sum_{k=0}^{\infty} \left[\frac{(\gamma^2/4)^{2k}}{k! \, (2k+1)!} \right] \bigg/ \frac{4}{\gamma^2} \sum_{k=1}^{\infty} \left[\frac{2k^2(\gamma^2/4)^{2k}}{k!^2(2k)!} \right] \right]$$

$$\approx R_{dc} \left[\frac{\gamma}{\sqrt{2}} + \frac{1}{2} + \frac{3}{8\sqrt{2}\gamma} \right] \qquad (\gamma > 10) \qquad\qquad (6.15)$$

$$G_{round} = \frac{-2\pi\gamma}{\sigma} \; \frac{(ber_2(\gamma) \, ber'(\gamma) + bei_2(\gamma) \, bei'(\gamma))}{ber^2(\gamma) + bei^2(\gamma)}$$

$$= \frac{2\pi\gamma}{\sigma} \left[\left(\sum_{k=1}^{\infty} \left[\frac{(-1)^k(\gamma/2)^{4k}}{(2k-1)!(2k-1)!} \right] \cdot \sum_{k=1}^{\infty} \left[\frac{(-1)^k(\gamma/2)^{4k-1}}{(2k-1)!(2k)!} \right] + \right. \right.$$

$$\left. \left. \sum_{k=0}^{\infty} \left[\frac{-(-1)^k(\gamma/2)^{2(2k+1)}}{(2k)!(2(k+1))!} \right] \cdot \sum_{k=1}^{\infty} \left[\frac{-(-1)^k(\gamma/2)^{4k-3}}{(2k-1)!(2k-2)!} \right] \right) \bigg/ \sum_{k=0}^{\infty} \left[\frac{(\gamma^2/4)^{2k}}{k!^2(2k)!} \right] \right]$$

$$\approx \frac{-\pi\gamma^4}{\sigma \, 8} \qquad (\gamma > 10) \qquad\qquad (6.16)$$

where $R_{dc} = \dfrac{4}{\pi d^2 \sigma}$

$\gamma = \dfrac{d}{\sigma \sqrt{2}}$

with d = diameter of the wire

6.3 EDDY LOSSES IN STRANDED WIRES

To reduce eddy current losses, particularly at higher frequencies, stranding of conductors can be used. Individual wires of the strand are enamelled and weaved along the entire divided conductor in such a way that all wires successively pass through all points of the cross-section. This ensures that the current will divide equally among the separate strands.

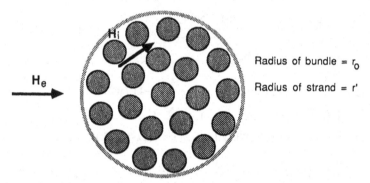

Radius of bundle = r_0

Radius of strand = r'

Figure 6.1 Cross section through a stranded conductor.

If there are a total of N strands, the skin effect losses is simply the sum of N conductors each carrying I/N of the currrent:

$$P_{1s} = \frac{F_{strand}\, I^2}{N} \tag{6.17}$$

Two magnetic field components cause proximity effect losses, namely an external field, H_e, and the internal collective field of the stranded conductor, H_i (see Figure 6.1). The internal magnetic field on radius r is given with sufficient accuracy by the first equation of Maxwell :

$$2\,\pi\,r\,H_i = \pi\,r^2\,J_0 \tag{6.18}$$

where the average current density J_0 equals;

$$J_0 = \frac{I}{N}\,\frac{p}{\pi\,r'^2} \tag{6.19}$$

with the packing factor p defined as follows:

$$p = \frac{N \pi r'^2}{\pi r_o^2} \tag{6.20}$$

From (6.18-20) we obtain:

$$H_i = \frac{I r}{2\pi r_o^2} \tag{6.21}$$

If θ is the angle between the vector and the x-axis, the internal magnetic field vector becomes:

$$H_i = H_i(\sin\theta x + \cos\theta\ y) \tag{6.22}$$

The total magnetic field equals the vector addition of H_i and H_e, with H_e lying on the x-axis:

$$H = (H_e + H_i \sin\theta)x + H_i \cos\theta\ y \tag{6.23}$$

The modulus of the square of the field intensity is:

$$H^2 = H_e^2 + H_i^2 + 2 H_e H_i \sin\theta \tag{6.24}$$

The total losses can be obtained by a tedious summation; instead the losses per unit cross section is found and then we proceed by integration instead of addition. The theoretical losses per unit area are:

$$S_1 = \frac{N}{\pi r_o^2} G_{strand} H^2 \tag{6.25}$$

The total proximity effect losses per unit length is obtained by integrating (6.25) over the cross section surface:

$$P_1 = \int_0^{r_o} \int_0^{2\pi} \frac{N\,G_{strand}}{\pi\,r_o^2} \left[H_e^2 + \frac{(I\,r)^2}{(2\,\pi\,r_o^2)^2} + \frac{H_e\,I\,\sin\theta}{2\,\pi\,r_o^2} \right] d\theta\; r\; dr$$

$$= N\,G_{strand} \left[H_e^2 + \frac{I^2}{8\,\pi^2 r_o^2} \right] \tag{6.26}$$

Equation (6.26) indicates that proximity effect losses due to an external field can be separated from the proximity effect losses imposed on the individual strands by neighbouring strands. Therefore:

$$P_{1p} = P_{1p\ external} + P_{1p\ internal} \tag{6.27}$$

$$\text{where} \quad P_{1p\ external} = N\,G_{strand}\,H_e^2 \tag{6.28}$$

$$P_{1p\ internal} = \frac{N G_{strand}\,I^2}{8\,\pi^2\,r_o^2} \tag{6.29}$$

6.4 CURRENT DISTRIBUTION ALONG THE WIDTH OF STRIP WINDINGS

One dimensional analysis assumes the current distribution along the width of a strip winding to be uniform. However, the current distribution in a strip winding is also determined by the radial flux density cutting it and may not be uniform. It is noteworthy that this issue is not addressed in papers on analytical methods, indicating lack of suitable methods to calculate this current distribution. Under conditions where breadth of the strip extends over the full window width and the packing factor of the winding is high, the error induced by ignoring the distribution is not large. It is nevertheless neccessary in comparative studies of possible configurations to be able to discern the influence of this on the total eddy losses. Mullineux[2] in 1969 identified this problem and presented a technique to do this, and applied it to the open side of a transformer consisting of a single wire wound winding and strip winding. However, even for this relatively simple configuration he had to resort to numerical integration methods.

Attempting to calculate the exact current distribution inevitably makes the problem a two dimensional one. Solving the eddy currents in an isolated rectangular conductor has been done in the literature[1]. The solution entails an infinite series, which does not converge very quickly and is rather lengthy to calculate. Unfortunately it ignores any proximity effect, limiting the scope of its application.

Another factor which should also be taken into account is the interaction between the source of the magnetic field and the induced circulating current due to proximity effect . The proximity effect, as described in paragraphs 4.5.3 and 4.5.5 assumes that the external field and its source is unaffected by the induced eddy currents. If it is to be included, it would typically entail partial magnetic coupling between the field source and induced eddy currents, which would result in increased losses. Ignoring the coupling is valid in cases where the current in the conductor under scrutiny has very little effect on the conglomerate magnetic field, for example one turn in a multi turn wire winding. This is not necessarily the case for the lateral current distribution in a strip conductor, because of the space occupied by the width of the conductor. In the example investigated by Mullineux, this coupling seems to be very small. Measurement on practical transformers wound with copper strips, also supports this notion, because such a magnetic coupling would induce considerable extra losses. On a heuristic level it seems that, because the major flux is carried in the magnetic cores for a transformer or inductor, very little parasitic magnetic coupling can occur in the stray field. Nevertheless, it can happen under certain conditions, e.g. a misaligned strip conductor in the fringe field of a gapped core.

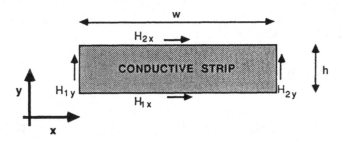

Fig. 6.2 Two dimensional analysis.

The approach to the two dimensional calculation of eddy current losses in strip conductors presented here, entails an approximate method where no magnetic coupling is assumed between the strip winding under scrutiny and other winding sections. The current distribution is taken to consist of two orthogonal one dimensional current distributions $J_x(x)$ and $J_y(y)$ so that;

$$I = I_x + I_y$$

$$= h \int_0^w J_x(x)dx + w \int_0^h J_y(y)dy \qquad (6.30)$$

The currents can also be expressed in terms of magnetic field intensity components (see Figure 6.2):

$$I_x = h(H_{2y} - H_{1y})$$

$$I_y = w(H_{2x} - H_{1x}) \qquad (6.31)$$

By inserting (4.38) and (4.39) into (4.49), an expression is found by which $J_x(x)$ and $J_y(y)$ can be solved:

$$J_a(a) = \alpha \left[(H_{2a} - H_{1a}) \frac{\cosh \alpha a}{\sinh \alpha b} - H_{1a} \tan \alpha b/2 \cosh \alpha a + H_{1a} \sin \alpha a \right] \qquad (6.32)$$

where a = x or y

       b = w or h

and α as defined in chapter 4.

The ohmic loss density is obtained from:

$$S_1(x,y) = \frac{1}{2\sigma} (J_x(x) + J_y(y)) (J_x^*(x) + J_y(y)^*) \qquad (6.33)$$

Integration of (6.32) over the rectangular cross section yields the total losses per unit length:

$$P_1 = \frac{1}{2\sigma} \int_0^h \int_0^w (J_x(x) + J_y(y)) (J_x^*(x) + J_y^*(y)) \, dxdy$$

$$= \frac{}{2\sigma} \int_0^w J_x(x)\, J_x^*(x)\, dx + \frac{}{2\sigma} \int_0^h J_y(y)\, J_y^*(y)\, dy \;+\; R_{dc}I_xI_y \qquad (6.34)$$

The total losses therefore reduces to two one dimensional eddy loss components of J_x and J_y and a frequency independent term $R_{dc}I_xI_y$. These eddy current loss components can be solved further by resolving it into terms for skin effect (6.9) and proximity effect (6.10).

A convenient method to solve the first term of (6.34), i.e. the lateral loss component of the strip, is by using the Poynting vector theory(see equation (2.1)). Applying the theory to the rectangular volume, one finds that the surfaces at the air-conductor interface have non zero Poynting vector values. Thus, the losses due to J_x are;

$$P_{1x} = \frac{}{2\sigma} \int_0^w J_x(x)\, J_x^*(x)\, dx$$

$$= \operatorname{Re}\left[\frac{}{\sigma} \left(J_x(w) H_{2y} - J_x(0)\, H_{1y} \right) \right] \qquad (6.35)$$

where J_x is calculated by using (6.32).

For the purpose of the computer algorithm, the Poynting vector technique is convenient to use, since the power density calculation along the width of the strip conductor already makes (6.32) available as a subroutine.

6.5 MANIFESTATION OF EDDY CURRENTS IN WINDING SECTIONS

In Chapter 2 it was indicated that power transfer inside a transformer is governed by the stray magnetic field. Equation (6.10), indicate an interesting analogy; in a transmission line the power transfer is VI and the losses I^2R, while in a transformer the densities are ExH and GH^2 respectively. In the energy flow model the electric field, $-\partial A/\partial t$, is inversely proportional to the radius and is for practical purposes not affected by the winding configuration, whereas the stray magnetic field intensity varies greatly inside the winding window and can be considered as the indicator of energy flow intensity associated with the winding configuration. It is therefore no coincidence that the proximity effect losses are proportional to the square of the magnetic field intensity, as this phenomenon can also be related in terms of the "obstacle in the energy flux path model" cited in Chapter 2.

We define a winding section for our purposes as a homogenously wound portion of a transformer, inductor or machine winding. Over its entire cross section the same wire or strip conductor is used and is uniformly packed. The losses in the j'th turn of the winding section, is obtained by adding skin and proximity effect losses (equations (6.9) and (6.10));

$$P_{j\ turn} = s_j\ (F\ I^2 + G\ H_{ej}{}^2) \qquad\qquad\qquad (6.36)$$

where s_j = the circumference of the j'th turn.

Note that I, F and G are constant over the winding section. In case of a monofilar winding all the turns carry a current I, while in the bifilar case centre tap transformer operation is assumed, where one half of the turns carry I and the other half zero current at any stage. Therefore the total skin effect losses inside a winding section is;

$$P_{skin} = N\ s_{av}\ F\ I^2 \qquad\qquad (monofilar) \qquad\qquad (6.37)$$

$$= \frac{N}{2}\ s_{av}\ F\ I^2 \qquad\qquad (bifilar) \qquad\qquad (6.38)$$

where s_{av} = average circumference of the winding section,

  N = number of turns in the winding section.

Since the skin effect loss is uniform over the cross section of the winding, the dissipation density is simply;

$$S_{skin} = \frac{P_{skin}}{A_{sect}} \qquad\qquad\qquad (6.39)$$

where A_{sect} = The cross section area of the winding section.

Similarly the power dissipation density due to proximity effect can be expressed as:

$$S_{proximity} = \frac{N\, s_{av}\, G\, H^2}{A_{sect}} \tag{6.40}$$

The total proximity effect loss can then be obtained by integrating the density over the cross section of the winding:

$$P_{proximity} = \frac{N\, s_{av}\, G}{A_{sect}} \int_{Asect} H^2\, dA \tag{6.41}$$

The total power dissipation density is found by adding (6.39) and (6.40):

$$S_{eddy} = \frac{N\, s_{av}}{A_{sec}} (F\, I^2 + G\, H^2) \qquad \text{(monofilar)} \tag{6.42}$$

$$= \frac{N\, s_{av}}{A_{sec}} \left(\frac{F\, I^2}{2} + G\, H^2 \right) \qquad \text{(bifilar)} \tag{6.43}$$

The total power dissipation is obtained through integration of (6.40) and adding it to (6.37,8):

$$P_{eddy} = N\, s_{av} \left(F\, I^2 + \frac{G}{A_{sec}} \int_{Asec} H^2\, dA \right) \qquad \text{(monofilar)} \tag{6.44}$$

$$= N\, s_{av} \left(\frac{F\,I^2}{2} + \frac{G}{A_{sec}} \int_{Asec} H^2\, dA \right) \qquad \text{(bifilar)} \tag{6.45}$$

The winding resistance at the i'th harmonic can be calculated from the ohmic loss equation:

$$R_i = \frac{2\,P_{eddy\,i}}{I_i^2} \qquad (6.46)$$

6.6 REFERENCES

[1] J Lammeraner, M Stafl ; "Eddy currents"; Iliffe Books - London; 1966.

[2] N. Mullineux, T.R. Reed, I.J. Whyte; "Current distribution in sheet and foil wound transformers", Proc IEE, Vol. 116; No 1; Jan 1969; pp. 127-129.

CHAPTER 7

MAGNETIC FIELD INSIDE WINDINGS; A KEY PARAMETER IN THE PROXIMITY EFFECT EQUATION

A detailed knowledge of the magnetic field intensity and distribution is required to solve the proximity effect losses. In the face of many numerical methods and algorithms in use today, the method of images was used. Two methods of modelling the windings are investigated; in the first case the winding is approximated as a uniform rectangular current distribution, neglecting the cylindrical curvature. In the second instance each turn in the winding is modelled by a circular current filament.

7.1 WINDING "STRAY" MAGNETIC FIELD

The proximity effect equations (6.40) and (6.41), consists of two product terms, the first term, G, is the same for every turn in a winding, and is a function of the conductor geometry, material and frequency. The other term is H^2, and is a funtion of position on the winding cross section. Calculation of the total proximity losses requires the square of the field intensity to be integrated over the cross section of the winding. Solving the magnetic field intensity crossing the winding conductors, which is often referred to as a "stray" field in literature, is therefore a very crucial part of any algorithm employing the equations of the previous chapter.

The technique which is selected to solve the stray field is very crucial, because it tends to be the most time consuming part of the algorithm. To simplify matters the assumption can in many cases be made that the ampere turns of a winding section is uniformly distributed over the cross section. This is not a serious limitation since, for most applications, large conductors are stranded or laminated to make the distribution as uniform as possible. This assumption simplifies the problem considerably and its solution is within the scope of a large number of techniques. In this age of numerical techniques, finite difference and finite elements are strong contenders. On the analytical side we have the classical method of images, Rogowski's method[3] using a single Fourier series, and Roth's method[2] which utilises a double Fourier series, which could all be viable alternatives. A finite element or difference method is very flexible and important aspects such as fringing around airgaps can be included. However, a difficulty of these methods is the definition of a mesh for each configuration, which would have to be preferably automatic for the CAD programs. Due to this and practical reasons, it was opted for the image method. It involves a two dimensional analysis which ignores the cylindrical curvature, which should not cause large errors for practical core configurations. As a matter of fact, Mullineux[5] claims that if the integral is taken, as we do when calculating the total losses, the effect is negligible.

7.2 METHOD OF IMAGES

(This paragraph is included for background to the reader, and is a short summary of Chapter 3 of reference [1].)

The method of images can be used to give solutions to problems involving straight line or circular boundaries in a particularly simple manner, for it offers certain ready made solutions, which eliminate the need for formal solutions of Laplace's and Poisson's equations. The essence of the method consists in replacing the effects of a boundary of an applied field by simple distributions of currents or charges behind the boundary line (called images). The desired field is then given by the sum of the applied and image fields.

When applying images to magnetic problems, the sign of the image has to be the same as the applied current, which gives only a normal magnetic field at the air magnetic material interface. The simplest case is the single boundary problem which involves only one image as indicated in Figure 7.1. A simplifying assumption, commonly made in magnetic problems, is that a boundary is of infinite permeability, which clearly involves some error. A convenient feature of the method of images is that it can accommodate finite permeability by adding a $(\mu_r-1)/(\mu_r+1)$ weighing factor to the image.

The solution becomes more complicated in the region between two parallel boundaries, because the number of images becomes infinite. (This effect can be observed by viewing one's images in parallel mirrors for example.) When a conductor is surrounded by four intersecting boundaries, the distribution of images is doubly periodic, and are symmetrical about each of the boundary lines as illustrated in Figure 7.2.

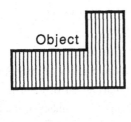

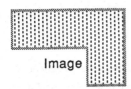

Figure 7.1 Single boundary problem.

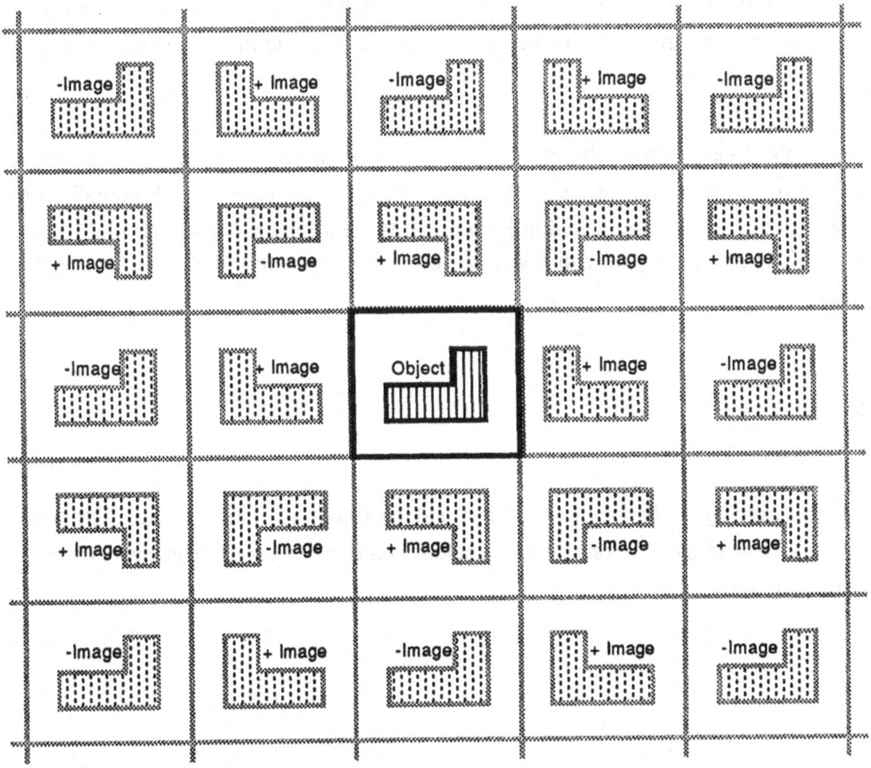

Figure 7.2 Four boundary problem.

7.3 METHOD OF IMAGES TO CALCULATE THE STRAY MAGNETIC FIELD IN WINDINGS

Strictly speaking the image method requires an infinite series of images for the plane boundries of the winding window. Since the contribution of each image must be calculated separately, which is very time consuming, the number of images has to be limited, thus allowing some error to creep is. In the algorithm of TID (Appendix C) only images which borders directly on the window are taken into account, as shown in Figure 7.3. Finite permeability of the core is being catered for, by adjusting the weight of the images by a factor, $(\mu_r - 1)/\mu_r$, as indicated.

The image method also lends itself to analise problems where magnetic materials are only present on one of the four winding window boundries (Figure 7.4), for example an inductor with a rod core (computer program CID, in Appendix D), or that portion of the winding circumference transformer where the windings are not enclosed by magnetic material (as employed in TID). The field calculation in this open window arrangement is also more accurate, since the image theory requires only one image and not an infinite number as in the previous case.

It is possible to get a reasonable approximation of the magnetic stray field in cases where one or two of the boundries are open ended, or airgaps are employed. It is realised by using the image system depicted in Figure 7.5. If the airgap g, is inserted in a yoke of length, l, and relative permeability μ_r, an effective permeability μ_r^* is calculated as follows:

$$\mu_r^* = \frac{l}{(g + (l-g)/\mu_r)} \qquad (7.1)$$

Note that fringing of the magnetic field around the airgap is not incorporated. The author is nevertheless of the opinion that this method should give reasonable answers for many practical designs.

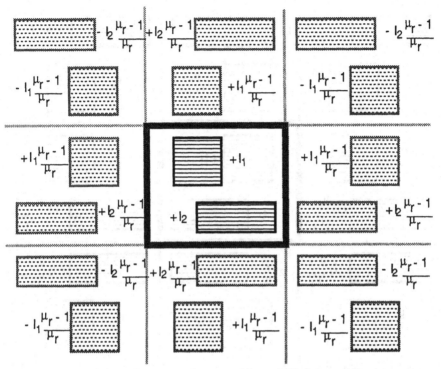

Figure 7.3 Mirror images for rectangular core with μ_r relative permeability.

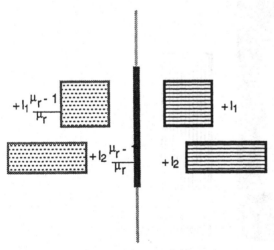

Figure 7.4 Single magnetic material boundary

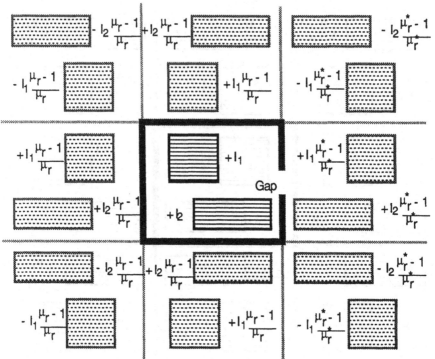

Figure 7.5 Window with airgap.

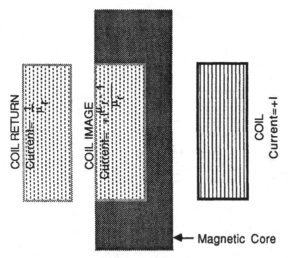

Figure 7.6 Calculation of the magnetic field in a coil in rectangular coordinates.

When the magnetic field of a air cored coil is calculated the influence of the return section must included. In order to accommodate magnetic and air cores in the same algorithm, a configuration involving two "images" is used for the calculation of the magnetic field inside the winding, as indicated in Figure 7.6. Firstly, the image $(\mu_r - 1)/\mu_r$ is the reflection of the winding in the magnetic material. The influence of the return half of the winding is given by the weighing factor $-1/\mu_r$. Thus for a highly permeable core the $(\mu_r - 1)/\mu_r$ image will dominate, whereas only the $-1/\mu_r$ term will play a role in the case of a air core inductor.

The image configuration of Figure 7.6 is applicable when a winding section is represented as a rectangular current distribution and the curvature ignored. The alternative field calculation as implemented in program CID, uses circular current filaments, and the the placement of an image is done[6] as shown in Figure 7.7. If the radius of the coil is R_c, that of the magnetic rod R_m, then the radius of the image will be R_m^2/R_c.

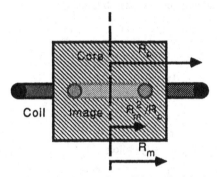

Figure 7.7 Image of a coil in cylindrical coordinates.

7.4 MAGNETIC FIELD OF A RECTANGULAR UNIFORM CURRENT DISTRIBUTION

Now that the placing of images has been defined, all that remains is to find an analytical expression with which the magnetic field can be calculated for a rectangular conductor carrying uniform current I. Such an expression is readily found in literature [1] and is as follows:

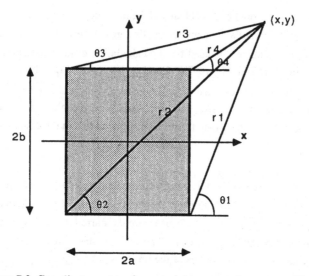

Figure 7.8 Coordinate system for a straight rectangular current distribution.

$$H_x = \frac{I}{8\pi ab}[(y+b)(\theta_1-\theta_2) - (y-b)(\theta_4-\theta_3) + (x+a)\log\frac{r_2}{r_3} - (x-a)\log\frac{r_1}{r_4}]$$

$$(7.2)$$

$$H_y = \frac{I}{8\pi ab}[(x+a)(\theta_2-\theta_3) - (x-a)(\theta_1-\theta_4) + (y+b)\log\frac{r_2}{r_1} - (y-b)\log\frac{r_3}{r_4}]$$

$$(7.3)$$

where $r_1 = \sqrt{(x-a)^2 + (y+b)^2}$ $\theta_1 = \arctan\frac{x-a}{y+b}$

$r_2 = \sqrt{(x+a)^2 + (y+b)^2}$ $\theta_2 = \arctan\frac{x+a}{y+b}$

$r_3 = \sqrt{(x+a)^2 + (y-b)^2}$ $\theta_3 = \arctan\frac{x+a}{y-b}$

$$r_4 = \sqrt{(x-a)^2 + (y-b)^2} \qquad \theta_4 = \arctan \frac{x-a}{y-b}$$

7.5 MAGNETIC FIELD OF CIRCULAR CURRENT FILAMENTS

Computer program CID (Appendix D) offers two methods to calculate the magnetic field at the windings. The one is the method of the previous paragraph, while the other is a more time consuming method which involves replacing the turns of the coil with circular current filaments; the topic of this paragraph.

To evaluate the magnetic field of a circular current at an arbitrary point in space tends to be a tedious process. The formulas for the magnetic field components of even a simple filamentary circular current are rather involved expressions of the coordinates, and of complete elliptic integrals of first and second kinds. The complication becomes more troublesome when the current is distributed such as in cylindrical current sheets, circular current disks or a multiple turns coil. The approach followed to calculate the magnetic field, is the one described in reference [4].

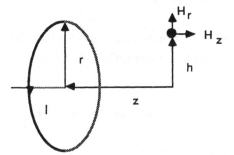

Figure 7.9 Coordinates of a point relative to a cylindrical current filament.

From the Bio Savart formula it can be shown [4] that the magnetic field components for a circular filamentary current in the coordinate system as defined in Figure 7.9 are:

$$H_r = \frac{-1}{2\pi} \int_0^\pi \frac{zr\cos\alpha \, d\alpha}{(h^2 + r^2 + z^2 + 2hr\cos\alpha)^{3/2}} \qquad (7.4)$$

$$H_z = \frac{1}{2\pi} \int_0^\pi \frac{(r2 - hr\cos\alpha)d\alpha}{(h^2 + r^2 + z^2 + 2hr\cos\alpha)^{3/2}} \qquad (7.5)$$

Since the integrals are very intractable for explicit solution they are being rewritten as summations which are solved numerically by the computer.

7.6 REFERENCES

[1] KJ Binns, PJ Lawrenson; "Analysis and computation of electric and magnetic field problems"; Pergamon Press - 1973.

[2] E Roth ; "Introduction a l'etude analytique de l'echauffement des machines electriques"; Bull. Soc. Franc. Elect., Vol 23, (1928), p773 .

[3] W Rogowski ; "Uber das Streufeld und der Streuinduktionskoeffizienten eines Transformators mit Scheibenwicklung und geteilten Endspulen"; Mitt. Forsch Arb. VDI71(1909).

[4] PJ Hart; "Universal Tables for Magnetic Fields of Distributed Circular Currents."; 1967, Elsevier - New York.

[5] N Mullineux, TR Reed, IJ Whyte; "Current distribution in sheet and foil wound transformers"; Proc IEE, Vol. 116; No 1; Jan 1969, pp. 127-129.

[6] JM Robertson; "Hydrodynamics in theory and application"; 1965, Prentice-Hall Inc.

CHAPTER 8

EXPERIMENTAL MEASUREMENT OF EDDY CURRENT LOSSES IN TRANSFORMER WINDINGS AND INDUCTOR COILS

The equations and theory of Chapters 6 and 7 are evaluated against experimental measurements described in the literature, and measurements conducted by the author himself. The outcome is promising but certain shortcomings, the most prominent being the calculation of the magnetic field, come to light.

8.1 METHODS OF MEASURING THE EFFECTIVE RESISTANCE OF INDUCTOR COILS

Accurate measurement of losses in coils and windings under waveform conditions, as encountered in high frequency switchmode circuits, seems to be beyond the scope of currently available wattmeters. The high frequency content of complex current and voltage waveforms, the very large reactive power or transferred power (in the case of transformers) to conduction losses ratio, as well as occurrences of high frequency parasitic resonances in the waveforms, push the operation of the wattmeters into an area where the measurement error can be larger than the actual signal. A particular limiting factor at high frequencies is the phase delay introduced by the current measurement[7]. Due to these factors Coonrod[6], for example, resorted to the old but proven thermal method of loss measurement.

Between 1910 and 1940 various experiments were conducted to determine the ac resistance of single layer coils. Many of these experiments utilised a resonant circuit to obtain ac current at the required frequency and thermal methods to determine the losses. An exception is Hickman, as reported by Butterworth[3], who used a bridge method at audio frequencies. Apart from the normal calorimetric methods, Wilmotte, as reported by Moullin[4], used a very elegant thermal method in which the coil of a resonant circuit consisted of a mercury spiral inside a glass tube and the thermal expansion of the mercury allowed the coil to act as its own thermometer.

Another method, also used during this period, is the resonant method, which measures essentially the quality factor of the ac circuit. A drawback of this method is that the measured resistance consists of the total circuit losses, which includes the coil, the capacitor and conductive leads. Dye[7] and Moullin[4] both proposed methods to separate the inductor and capacitor losses. Contrary to Dye's method, the scheme proposed by Moullin not only measures the capacitor losses, but also any other losses not associated with the coil. An additional advantage of Moullin's method is its simplicity, perhaps one of reasons why Jackson[5] applied it to test Butterworth's equations.

8.2 MEASUREMENTS ON SINGLE LAYER COILS AND COMPARISON OF ALGORITHM WITH BUTTERWORTH'S EQUATIONS

8.2.1 The classic work by Butterworth

The yardstick to evaluate an algorithm which predicts ac resistance of single layer coils is undoubtedly the work by Butterworth. In 1921 S. Butterworth published a classic contribution on the high frequency resistance of short single layer coils[1]; this was followed by two other papers[2,3], which extended the work to coils of any length and any number of turns. In his first paper[1] Butterworth calculated the high frequency of a band of similar parallel wires, each of diameter d and with their axes separated by a distance D. He showed that the effective resistance of a wire depended on its position in the band but that its average resistance R_f at frequency f was related to R_o , the direct current resistance per unit length, by the equation;

$$\frac{R_f}{R_o} = 1 + F(z) + \frac{\mu_n d^2 \, G(z)}{D^2} \tag{8.1}$$

where μ_n, $F(z)$ and $G(z)$ are complicated functions which he tabulated.

In his third paper[3] he made a notable extension to the above equation in the form of the following equation, which is applicable to coils having "an extremely large number of turns";

$$\frac{R_f}{R_o} = \alpha(1 + F(z)) + \frac{G(z)\ (\beta\mu_1 + \gamma\mu_2)\ d^2}{D^2} \tag{8.2}$$

where α, β, γ and μ_1 and μ_2 are tabulated functions.

Section (10) of this paper deals with the resistance of coils having a finite number of turns, but these are to be very short: for such coils he finds that;

$$\frac{R_f}{R_o} = 1 + F(z)\left(1 + \frac{\omega_n d^4}{8D^4}\right) + \frac{\mu_n G(z)\ d^2}{D^2}\left(1 + \frac{v_n\phi_1 d^2}{2D^2}\right) \tag{8.3}$$

where v_n, ω_n, and ϕ_1 are tabulated functions.

8.2.2 Comparison between theoretical predictions and experimental measurements

Predictions by the equations in this paper are plotted against the observed values and Butterworth's predictions in Figures 8.1 to 8.5. In these comparisons the following algorithms are compared to the measured values:

a) Magnetic field calculated by assuming a uniformly distributed current across the winding and rectangular coordinates - referred to as "rectangular" in graphs.

b) Magnetic field calculated by replacing individual turns by current filaments - referred to as "cylindrical" in graphs.

c) Formula 8.2 - referred to as "Butterworth 2" in graphs.

d) Formula 8.3 - referred to as "Butterworth 3" in graphs.

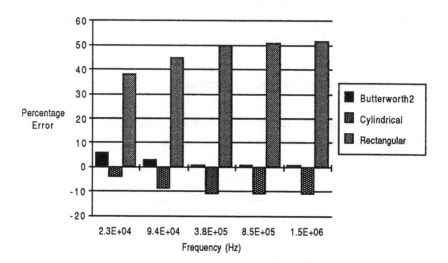

Figure 8.1   COIL 1; Howe's measurements[3] against predictions.
              Wire Diameter = 1.63 mm
              Length of coil = 206 mm
              Coil Diameter = 20.6 mm
              Number of turns = 62

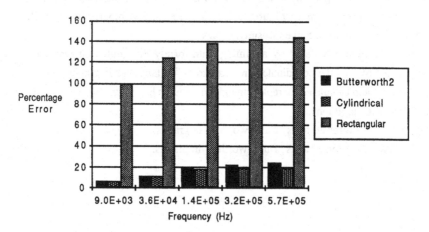

Figure 8.2   COIL 2; Howe's measurements[3] against predictions.
              Wire Diameter = 2.64 mm
              Length of coil = 269 mm
              Coil Diameter = 26.9 mm
              Number of turns = 92

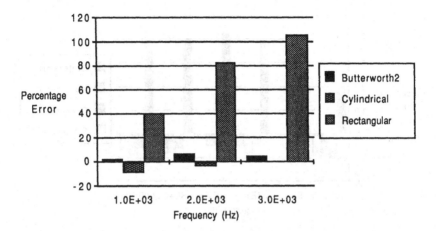

Figure 8.3   COIL 3; Hickman's measurements[3] against predictions with
coil diameter = 82.4 mm.

Wire Diameter = 5.18 mm

Length of coil = 960 mm

Number of turns = 160

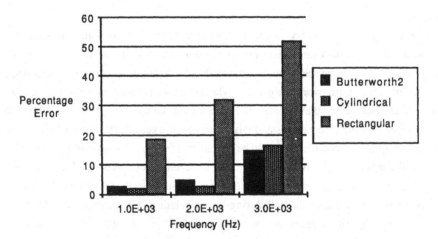

Figure 8.4   COIL 4; Hickman's measurements[3] against predictions with
coil diameter = 304 mm.

Wire Diameter = 5.18 mm

Length of coil = 960 mm

Number of turns = 160

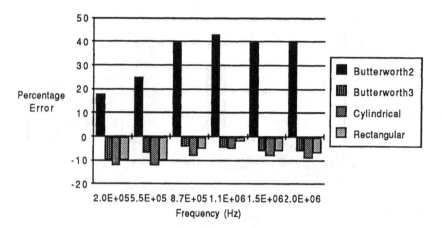

Figure 8.5   COIL 5; Jackson's observations[5] against predictions.
                Wire Diameter = 1.63 mm
                Length of coil = 23.4 mm
                Coil Diameter = 102 mm
                Number of turns = 10

Discussion

Howe did experiments on very long solenoids and varied the ratio of diameter to distance
between turns. Hickman varied the length to diameter ratio, while Jackson investigated a
very short coil. It can be seen from the graphs that the "cylindrical" algorithm performed
almost as well as Butterworth's equations. The deviations from the measured values can
be attributed to the fact that the current distribution is not uniform across the conductor
cross sections at high frequencies and this can have a marked effect on the field in single
layer solenoids. A second reason could be the fact that for the calculation of the proximity
effect, the magnetic field is uniform across a conductor cross section, which is a good
approximation for a multi-layered structure, but not necessarily for a single layer coil.
Ignoring the curvature of the coil during the calculation of the magnetic field in the
"rectangular" algorithm, tends to give overly optimistic predictions on the ac resistance
for long solenoids, but performs well in the case of short coils and is useful when the
length to diameter ratio is smaller than one.

8.3 MEASUREMENT OF MULTILAYER COILS AND TRANSFORMER WINDINGS

8.3.1 Principle of the experimental method

The experimental method is essentially a modern version of the measuring technique, as described by Moullin[4] and Jackson[5], utilising a digital oscilloscope and computer to acquire and process the data. The method measures the natural responce of a series RLC circuit, and finds the best analytical function to fit the measured waveform, from which the circuit resistance and inductance at the specific frequency is calculated. The details of the experimental procedure and the measured data can be found in Appendix A. In this paragraph the principle of the method to extract the ac residual resistance from the measured data is described.

The total resistance of a resonant circuit is measured when the inductance is a coil of copper wire and again when this inductance has been replaced by an identical coil of brass or german silver. In substituting one coil for another, the only change in the circuit is the conductivity of the coil conductors, and therefore any losses that are not located in the coil remain unchanged by the substitution. At a given frequency the total resistance of the circuit may be expected to consist of a constant component due to the combined losses in the capacitor, conductor leads, eddy currents induced in metal parts of the structure other than the coil itself, and even the contact resistance of the coil circuit connections; and a second component which is a function of the conductivity of the coil material.

If R_c is the calculated resistance of the coil at a given frequency, then R_t, the total measured resistance of the circuit may be expected to be of the form;

$$R_t = R_r + \kappa R_c \qquad\qquad (8.4)$$

where R_r represents the effective residual resistance and κ is a factor which should approach one if the model or formula being used is accurate. Since the relation between R_t and R_c should be a straight line, R_r will be determined by the intercept on the total resistance axis, and its value determines the resistance the circuit would have if the coil could be made of a material having a infinite conductivity. Hence by plotting the predicted resistance against the measured values for coils with identical construction but different materials, the validity of the computing algorithm or formulae as well as the experimental method can be checked, by noting that the value of κ and the scattering of data points on the graph.

This method of procedure is taken one step further in this work. Firstly, the short coil construct of Jackson[5] has been duplicated for the conductive materials copper, brass and german silver. These coils are used to find the residual resistance at various frequency points. By interpolating between the residual resistance values at several measured frequency points the circuit can be effectively calibrated over the frequency range, thus making it possible to measure the ac resistance of any unknown coil which has an inductance close to that of the calibration coils. Futhermore, by measuring the ac resistance of the circuit when a transformer with a shorted secondary is included in the resonant circuit, and subtracting it from the ac resistance measurement when the transformer is removed from the circuit, it is possible to measure the primary referred ac resistance of the transformer.

8.3.2 Comparison between theoretical prediction and measurement on air coils

The calibration procedure, as described in Appendix A, yielded a determination coefficient, for a straight line fit, that is larger than or equal to 99.9% and a slope that deviates by less than 4% from unity. The calibration therefore supports the validity of the algorithm for the specific single layer coils, and the experimental method to an accuracy of roughly 5%. The residual resistance at the measured frequencies is shown in Figure 8.6. R_r reaches a minimum at 17.6 kHz. At lower frequencies the increase can be attributed to the characteristics of the semiconductor switch, whereas at higher frequencies the skin and proximity effect in the circuit plays a major role.

The ac resistance at seven different frequency points was measured for each of six coils. The first three are the single layer "Jackson" coils, made of copper, brass and german silver, wire diameter ranging between 1.6 and 1.7 mm. The others are multilayer coils, each composed of three layers of five turns each, using different copper conductors of solid wire, litz wire, and flat strip. In the following six bar graphs the predicted and measured resistance values are displayed.

Residue Resistance (mΩ)

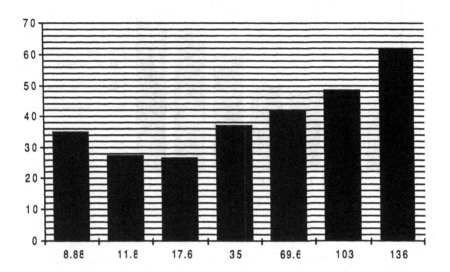

Figure 8.6 The residual resistance as a function of frequency.

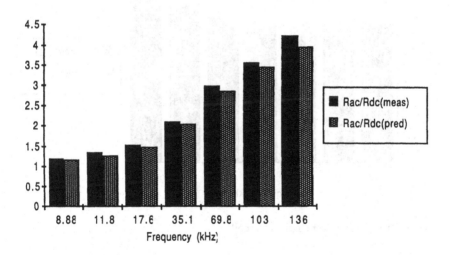

Figure 8.7 Copper single layer coil.

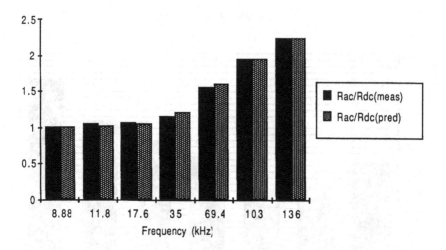

Figure 8.8 Brass single layer coil.

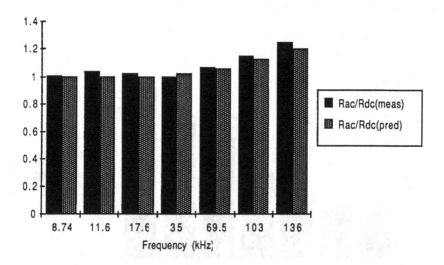

Figure 8.9 German silver single layer coil.

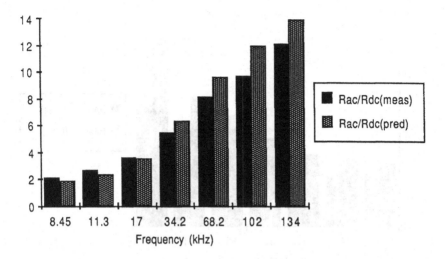

Figure 8.10 Multi layer solid copper wire coil.

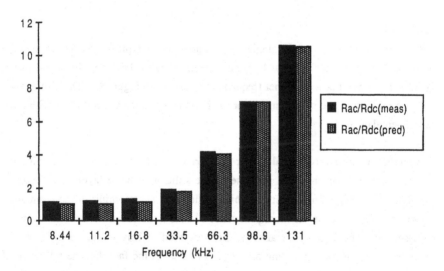

Figure 8.11 Multi layer litz copper wire coil.

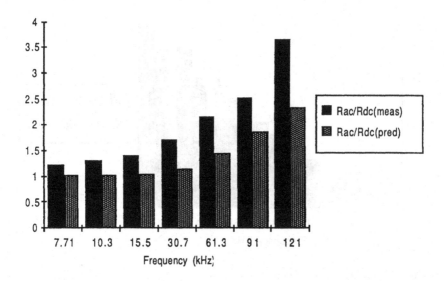

Figure 8.12   Multi layer copper strip conductor coil.

Discussion

The predicted resistance of the single layer copper coil is typically 5% lower than the measured value, which is in fairly good agreement with the prediction error in the Jackson experiment done at higher frequencies as shown in Figure 8.5. The discrepancy between prediction and measurement for the brass and german silver coils is within the experimental error.

The prediction error for the multi layer litz wire coil is fairly acceptable, keeping the experimental error in mind. In the case of the solid wire multi layer coil, the error increases noticebly with frequency, which indicates a shortcoming in the computing algorithm. This tendency is more pronounced in the transformer measurements in the next section. The best explanation to be presented at this stage, keeping all the approximations of the computing algorithm in mind, is the fact that the influence of proximity effect induced eddy currents in windings is not taken into account during the calculation of the magnetic field. Since the eddy currents will reduce the magnetic field, the proximity effect loss will be reduced, which in turn will decrease the prediction error for the solid wire coil, and to a lesser extent for the litz wire.

The prediction error increases to beyond thirty percent for the strip conductor coil. It is felt that a major cause of this error should be ascribed to the fact that the conductors are so widely spaced, and this can lead to a magnetic field which is not predominantly parallel to the strip conductor surface. The model for calculation of losses in strip conductors is not the exact analytic two dimensional solution, but a quasi two dimensional method that requires the parallel field to dominate the field perpendicular to the strip conductor.

8.3.3 Comparison between theoretical predictions and measurements on transformer windings

The method of measurement, described in more detail in Appendix A, is essentially a short circuit test with the primary being connected in series with the resonant circuit. The calibration was done by removing the transformer from the circuit, and shorting the circuit terminals provided for the transformer. A relatively large frequency difference between the calibration and transformer measurements was observed due to the leakage inductance of the transformer, which required that frequency compensation had to be done on the calibration resistance measurement before it could be subtracted from the transformer measurement. This compensation, combined with the non-linear characteristics of the semiconductor, and the fact that the curve fitting algorithm was less refined during the transformer measurements than when measurements were done on coils, causes a larger potential experimental error, which could be as high as an estimated 20%.

In the following figures the predicted and measured primary referred resistance as a function of frequency is displayed. All the transformer configurations share the same secondary winding, which is a four turn strip winding, shorted at the terminals. The primary winding configuration covers a spectrum of possibilities: solid wire, litz wire, strip conductors and windings that fill the full window width, half of the width, and a winding that is bifilar wound.

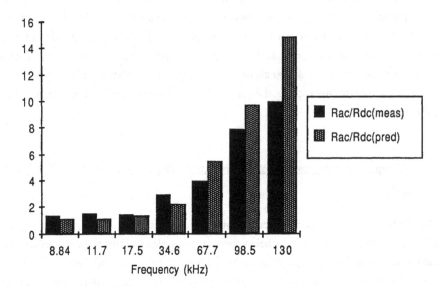

Figure 8.13 Full height primary strip winding.

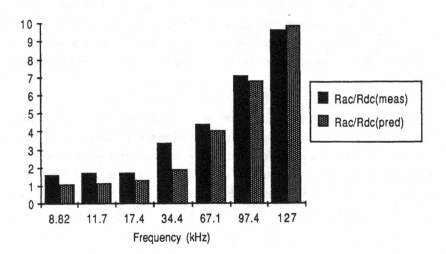

Figure 8.14 Half height primary strip winding.

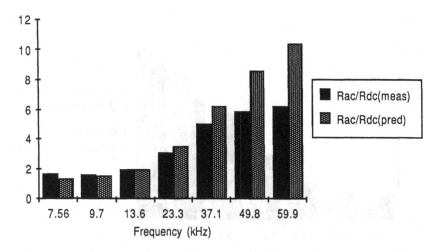

Figure 8.15 Full height solid wire winding.

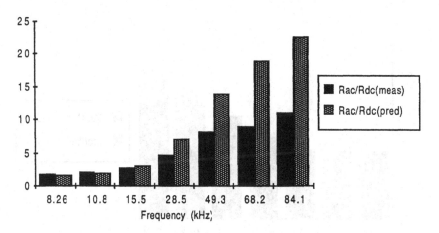

Figure 8.16 One half height primary solid wire active winding, other half winding passive.

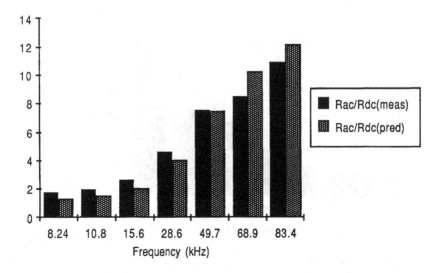

Figure 8.17 Half height primary solid wire winding.

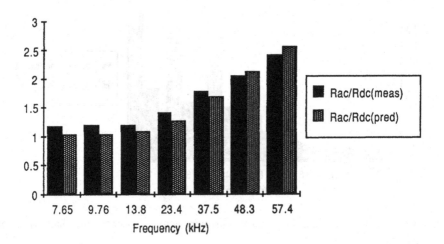

Figure 8.18 Full height primary litz wire winding.

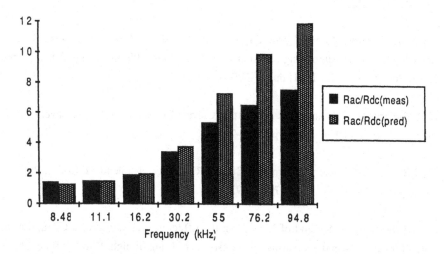

Figure 8.19 Full height primary bifilar solid wire winding.

Discussion

The predictions are fairly reasonable, taking the experimental uncertainty into account. The biggest source of error, larger than in the case of the coils, is the influence of the proximity effect induced currents on the magnetic field distribution inside the winding window. Probably the best example of how the algorithm overpredicts at high frequencies can be seen in the graph of Figure 8.16, when a passive winding is present, and comparing it to Figure 8.17 when it is absent. As with coil measurements the best correlation between measurement and prediction is obtained with litz wire because the proximity effect is small due to the thin strands.

Another observation, common to almost all the measurements, is that the predictions tend to be pessimistic at low frequencies. This tendency, though not then as prominent, is also observed during the coil measurements. The reason for this is not very clear at this stage. This deviation could on the one hand be within the experimental error; the non-linear characteristics of the semiconductor switch would probably be the biggest culprit. In such a case an improvement may be obtained by replacing the semiconductor switch with a mercury wetted relay. On the other hand, it may be a second order effect that the algorithm does not account for, such as eddy currents due to external stray fields.

8.4 REFERENCES

[1] S Butterworth; "Eddy-Current Losses in Cylindrical Conductors with Special Applications to the Alternating Resistance of Short Coils"; Philosophical Trans. of the Royal Society, vol 222, 1921, pp. 57-100.

[2] S Butterworth; " Note on the Alternating Current Resistance of Single Layer Coils"; Physical Review, 1924, vol 23, pp. 752-755.

[3] S Butterworth; "On the Alternating Current Resistance of Solenoidal Coils"; Proc. of the Royal Society, 1925, vol 107, pp. 693-715.

[4] EB Moullin; "A Method of Measuring the Effective Resistance of a Condenser at Radio Frequencies, and Measuring the Resistance of Long Straight Wires."; Proc. of the Royal Society, A , 1932, vol 137, pp. 116-133.

[5] W Jackson; "Measurement of the High Frequency Resistance of Single Layer Solenoids."; Inst. of Elec. Eng. Journal, April 1937, vol 80(484), pp. 440-445.

[6] NR Coonrod; "Transformer Computer Design Aid for Higher Frequency Switching Power Supplies."; IEEE Trans. on Power Electronics, vol PE-1, no. 4, Oct 1986, pp 248-256.

[7] JA Ferreira; "Experimental Evaluation of Losses in Magnetic Components for Power Converters.", Conference Record, 1988 IEEE IAS Annual Meeting, pp 668-673.

CHAPTER 9

EVALUATION AND CONCLUSIONS

This work had its roots in the belief that a deeper understanding of the electromagnetic power conditioning mechanism is necessary in order to keep pace with the trend towards an ever increasing frequency-power product in switchmode converter technology. Circuit design with the objective of keeping skin effect and proximity effect losses a minimum has been identified as a major criterion in the optimisation of high frequency converters. Overall, four major contributions emerged from this effort: firstly, energy flow modelling as a means of gaining insight into the mechanism of power conditioning; secondly, an algorithm has been devised to analyse structural impedances; thirdly, a new set of equations have been derived to model skin and proximity effect losses in magnetic components; and lastly, an experimental method has been devised to measure the ac resistance of magnetic components.

9.1 MODELLING OF ENERGY FLOW WITH THE POYNTING VECTOR

In power electronic converters, the prime concern is power conditioning and conversion. A current and voltage description, from this viewpoint, is an indirect way of modelling these objectives, whereas the Poynting vector provides a direct method of modelling energy flow. In the past Poynting vector theory has not been utilised to its full potential for describing energy flow in circuits, and this book focuses on the various mechanisms of energy flow in different circuit elements. In this age of sophisticated numerical techniques, it should be possible to implement energy flow analysis on a CAD workstation, thus making energy flow modelling accessible to engineers.

It is envisaged that an energy flow approach, supplementing conventional circuit analysis, can become an invaluable design tool. Such an energy flow analysis is, contrary to circuit analysis, a three dimensional model, which inevitably gives more insight into a problem. A valuable feature is the fact that eddy current losses in connection structures and magnetic components can be regarded as being a consequence of energy flow around

conductive structures. A Poynting vector approach can therefore be very useful during
the design of specialised magnetic components and the layout of high frequency circuits.

9.2 STRUCTURAL IMPEDANCE OF POWER CONNECTIONS

The approach to the problem of structural impedances has been to view a power
connection as a short transmission line. Arising from the work is the notion that a
connection is not always inductive, but can under certain conditions be capacitive or even
reactance free. On the analytical side, an algorithm has been developed and implemented
in a computer program, that allows a circuit designer to compute structural impedance
values. The formulae of Wheeler[reference 4,chapter 5] have been used to compute the
external impedance, whereas a first order model combining several analytical functions
which are easy to compute, is being used for the internal impedance calculation. The
algorithm has not been subjected to extensive experimental verification, as in the case of
the transformer and inductor algorithm. The only comparison against measurements was
the measurements on parallel wires as reported by Butterworth[reference7, chapter 5],
which indicated that the algorithm gives reasonable predictions for that particular
configuration. The computer program is, despite some shortcomings, a valuable tool to
get an indication of what the structural impedances for a specific design would be.

9.3 FORMULAE TO CALCULATE CONDUCTOR LOSSES IN MAGNETIC COMPONENTS

A set of analytical expressions has been derived which provides a means to analyse the ac
resistance and losses in the windings of magnetic components. These equations are very
versatile in that any periodic current waveshape and any winding or coil of solid wire, litz
wire or strip conductors can be accommodated. The occurrences of orthogonality in the
power integrals decouples the essential dissipation sources and frequency components,
which simplify the computation algorithm.

The tendency in many present day papers on transformers[3, 4, 5, reference 6 chapter 8]
for power electronic applications seems to be to use the one dimensional equations of
Dowell[reference 11, chapter 4] to analyse conduction losses. The one dimensional
solution, however, which excludes the effects of the magnetising current component in
the transformer, is generally not applicable to inductors and is subject to certain
configurational restrictions. The solutions presented in this thesis are two dimensional

and are not subject to these restrictions. As a matter of fact, Dowell's equations are included in a somewhat different format as sub-equations in the algorithm for strip conductors.

The experimental measurements seem to support the validity of these equations. The predictions are comparable to those of Butterworth's formulae for single layer coils. The major cause for deviations from the measurements on multi layer coils and transformers seems to be the influence of the proximity effect induced eddy currents on the magnetic field, which has not been included in the computer algorithms. Despite this, the predictions are reasonably accurate, and bearing in mind that nothing comparable is available at the time of writing, they are a useful tool for optimising conductor losses in magnetic components.

The effect where the induced eddy currents alter the magnetic field distribution involves the effective average current distribution across a complete winding section or coil. On the level of individual conductors the reaction of eddy currents on the magnetic field has already been incorporated during the derivation of the equations. As a matter of fact, an approach which ignores the influence of the eddy currents on the magnetic field inside the conductors has been described by Kaul[1]. Since the equations of Kaul are simpler and more flexible, the present author initially started with a computer algorithm based on his work[2]. However, when it became obvious that under certain conditions the predictions will grossly overestimate the losses, this approach was abandoned.

9.4 EXPERIMENTAL PROCEDURE TO MEASURE AC RESISTANCE

By reviving Moullin's method, developing it to its full consequences and using modern measuring instruments, an effective method has been devised to measure the ac resistance of coils and transformers. It is believed that the experimental error can be further reduced by making some hardware modifications to the circuit. Firstly, the non-linear on resistance of the semiconductor switch is a source of possible experimental error; replacing it with a mercury wetted relay will enhance the accuracy. Other improvements could be to reduce the number of mechanical contacts in the circuit, and to get better resolution during digitisation of the measured waveform.

The measuring procedure, in combination with the derived eddy current equations, has possibilities which have not been exploited fully during the experimental work of this thesis. Firstly, it can be utilised to measure the leakage inductance of transformers and the nominal inductance of coils as a function of frequency. Secondly, measurements can be done on the structural impedance of connection leads (as described in Chapter 5), if a sufficiently long lead of homogeneous construction is inserted in the circuit. Thirdly, since the parameters F and G, as defined in equations (6.11-16), are functions of the ratio of the conductor dimensions to depth of penetration, it is possible to do scaling of magnetic components. For example, it should be possible to test a scale model of a 10 kVA 50 Hz transformer, which could be a small transformer rated at perhaps 500 VA and 1 kHz. Since a short circuit test is conducted, the core losses are very small and it might not even be necessary to use special high frequency core material for the scale model.

9.5 REFERENCES

[1] HJ Kaul ; "Stray-current losses in stranded windings of transformers."; Trans. of AIEE, 76(30) 1957, pp137-149.

[2] JA Ferreira ; "Tweedimensionele analise van transformator/induktor-wikkelings, deel 1: die oplos van parasitere strome."; RAU Internal Report - END-33.

[3] B Carsten ; "High frequency conductor losses in switchmode magnetics"; PCI (Power Conversion International) June 1986 Proceedings (Munich), pp161-182.

[4] JP Vandelac, P Ziogas ; "A novel approach to high frequency transformer copper losses"; 1987 PESC Conference Record (Blacksburgh), pp355-367.

[5] L Bonte, J van Campenhout ; "A simplified high frequency network presentation of power pulse transformers for switched mode dc-dc converters and dc-ac inverters"; 1985 EPE Conference Record (Brussel),pp1.35-1.42.

APPENDIX A

EXPERIMENT FOR DETERMINING THE AC RESISTANCE OF INDUCTOR COILS AND TRANSFORMER WINDINGS

A.1 EXPERIMENTAL METHOD

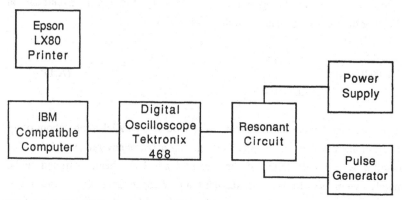

Figure A.1 Block diagram of experimental hardware.

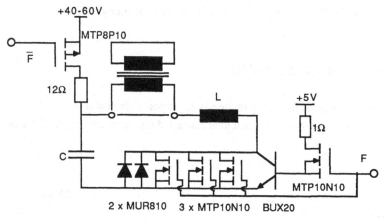

Figure A.2 Circuit diagram.

The experimental setup, as indicated in Figure A.1, consists of the resonant circuit, an oscilloscope that measures and digitises the waveform and a computer that does a curve fitting to the waveform to determine the ac resistance.

The circuit is a series RLC, with the resistance being the internal resistance of all the circuit components. Note that the semiconductor switch consists of a parallel combination of a bipolar transistor, two diodes and three MOSFET transistors, in order to keep the series resistance of the switch as small as possible. A transformer with a shorted secondary can be optionally inserted, thus adding the primary referred resistance and leakage inductance to the existing circuit series inductance and resistance. The capacitor is charged by turning the top transistor on and the bottom transistor off. Then by inverting the transistor on/off states the energy in the capacitor decays through the circuit natural response, giving the following capacitor voltage waveshape:

$$V(t) = V_0 \, e^{-tR/2L} \cos\left(\sqrt{\left[\frac{1}{LC} - \frac{R^2}{4L^2} \right]} \, t \right) \hspace{2cm} (A.1)$$

The inductor and capacitor of the circuit consists of serial and parallel combinations of up to four 14μH inductors and 1.5μF capacitors, giving a frequency range of 1:16. The stray capacitance and inductance of the circuit leads have been designed to be approximately two orders of magnitude smaller than the capacitance of the capacitor or coil inductance. The resistance on the other hand, consists of the collective value of all the circuit components. The major difficulty of an experiment of this type is to extract the resistance of the magnetic components of interest from the total circuit resistance.

A.2 EXPERIMENTAL MEASUREMENT

If the observed damping and frequency of the natural response are α and ω respectively, and if the capacitance is known, the circuit inductance and resistance can be calculated as follows:

$$L = \frac{1}{C(\omega^2 + \alpha^2)} \hspace{2cm} (A.2)$$

$$R = 2L\alpha \tag{A.3}$$

The capacitance was measured accurately to three decimal places on a capacitance bridge, while the damping and frequency is determined from the best fit to the experimental data of the analytical function (A.1), computed with the program described in Section A.5 of this appendix. Seven frequency points were measured. The lowest value was obtained when the configuration consisted of four parallel capacitors and four series inductors, while at the highest frequency the four capacitors were in series and the inductors all in parallel. At each frequency value six different coils with approximately the same inductance values, a transformer calibration and seven different transformer configurations were measured.

Coil Configurations:
(See Figure D.1 for dimension definitions)

SINGLE LAYER COPPER WIRE COIL (SCC)
    Wire diameter = 1.6 mm
    Average coil diameter = 100 mm
    Coil length = 27.5 mm
    Number of turns = 1x10
    DC resistance = 27.5 mΩ

SINGLE LAYER BRASS WIRE COIL (SBC)
    Wire diameter = 1.7 mm
    Average coil diameter = 100 mm
    Coil length = 27.5 mm
    Number of turns = 1x10
    DC resistance = 93.2 mΩ

SINGLE LAYER GERMAN SILVER WIRE COIL (SGSC)
    Wire diameter = 1.6 mm
    Average coil diameter = 100 mm
    Coil length = 27.5 mm
    Number of turns = 1x10
    DC resistance = 368.4 mΩ

MULTI LAYER SOLID COPPER WIRE COIL (MWC)

Wire diameter = 1.5 mm
Average coildiameter = 49 mm
Coil length = 9 mm
Coil height = 5 mm
Number of turns = 3x5
DC resistance = 23.1 mΩ

MULTI LAYER LITZ COPPER WIRE COIL (MLC)

Strand diameter = 0.5 mm
Number of strands = 6
Average coil diameter = 49 mm
Coil length = 9 mm
Coil height = 5 mm
Number of turns = 3x5
DC resistance = 35.1 mΩ

MULTI LAYER STRIP CONDUCTOR COIL (MSC)

Conductor width = 2.8 mm
Conductor thickness= 0.33 mm
Average coil diameter = 70 mm
Coil length = 18 mm
Coil height = 8 mm
Number of turns = 3x5
DC resistance = 62.6 mΩ

Transformer Configurations:

GENERAL

Core type = RM14
Core material = N67 ferrite
Conductor material = copper
Secondary is shorted and common to all the configurations

SECONDARY STRIP CONDUCTOR WINDING

Conductor width = 18 mm
Conductor thickness= 0.30 mm

Inner winding diameter = 16 mm
Winding height = 2.5 mm
Number of turns = 4x1
DC resistance = 1.01 mΩ
Connection lead width = 9 mm

FULL HEIGHT STRIP CONDUCTOR PRIMARY WINDING (1SW)

Conductor width = 18 mm
Conductor thickness= 0.30 mm
Inner winding diameter = 22 mm
Winding height = 3.25 mm
Number of turns = 6x1
DC resistance = 1.88 mΩ
Connection lead width = 9 mm

HALF HEIGHT STRIP CONDUCTOR PRIMARY WINDING ($^1/_2$SW)

Conductor width = 9 mm
Conductor thickness= 0.30 mm
Inner winding diameter = 22 mm
Winding height = 3.25 mm
Number of turns = 6x1
DC resistance = 5.62 mΩ
Connection lead width = 9 mm

FULL HEIGHT SOLID WIRE PRIMARY WINDING (1WW)

Wire diameter = 1.0 mm
Winding width = 19 mm
Inner winding diameter = 22 mm
Winding height = 3.25 mm
Number of turns = 3x14
DC resistance = 74.28 mΩ

HALF HEIGHT SOLID WIRE PRIMARY WINDING AND HALF HEIGHT PASSIVE SOLID WIRE WINDING ($^1/_2$WW$^1/_2$P)

Active and passive windings are identical
Wire diameter = 1.0 mm
Winding width = 9 mm
Inner winding diameter = 22 mm

Winding height = 3.25 mm
Number of turns = 3x7
DC resistance = 37.14 mΩ

HALF HEIGHT SOLID WIRE PRIMARY WINDING ($^1/_2$WW)

Wire diameter = 1.0 mm
Winding width = 9 mm
Inner winding diameter = 22 mm
Winding height = 3.25 mm
Number of turns = 3x7
DC resistance = 37.14 mΩ

FULL HEIGHT LITZ WIRE PRIMARY WINDING (1LW)

Strand diameter = 0.31 mm
Number of strands = 6
Winding width = 19 mm
Inner winding diameter = 22 mm
Winding height = 3.25 mm
Number of turns = 3x14
DC resistance = 168.4 mΩ

FULL HEIGHT LITZ BIFILAR PRIMARY WINDING (1BWW)

Wirediameter = 0.31 mm
Winding width = 17.5 mm
Inner winding diameter = 22 mm
Winding height = 3.5 mm
Number of turns = two bifilar 3x7
DC resistance = 37.36 mΩ

Measurements:

Due to the nature of these experiments, the measurements have statistical distributions associated with them. The experimental procedure was conducted as follows:

- A warm up period of between five and ten minutes was allowed for each configuration.

- A series of measurements are done, and if the average doesn't move, a set of six measurements are taken of which the average and sample standard deviation are calculated.
- An average that shifts was nomally caused by either bad circuit connections or insufficient warm up time.

The inductors and capacitors were connected as follows at the seven frequency points:

frequency 1 (f1)	4 parallel capacitors
	4 series inductors
frequency 2 (f2)	3 parallel capacitors
	3 series inductors
frequency 3 (f3)	2 parallel capacitors
	2 series inductors
frequency 4 (f4)	1 capacitor
	1 inductor
frequency 4a (f4a)	2 parallel capacitors
	2 parallel inductors
frequency 4 (f4b)	2 series capacitors
	2 series inductors
frequency 5 (f5)	2 series capacitors
	2 parallel inductors
frequency 6 (f6)	3 series capacitors
	3 parallel inductors
frequency 7 (f7)	4 series capacitors
	4 parallel inductors

COIL TYPE		Frequency (kHz)		Circuit Resistance (mΩ)	
		Average	Sample standard deviation	Average	Sample standard deviation
SCC	f1	8.877	0.001	165.5	0.3
	f2	11.77	0.04	138.4	0.9
	f3	17.64	0.01	110.3	0.2
	f4	35.17	0.02	95.11	1.25
	f4a	35.01	0.01	65.82	0.34
	f4b	35.28	0.01	156.2	0.1
	f5	69.80	0.01	89.76	0.22

	f6	103.1	0.1	94.73	0.12
	f7	136.2	0.1	110.3	1.5
SBC	f1	8.883	0.027	416.1	1.5
	f2	11.80	0.01	323.2	1.5
	f3	17.61	0.01	227.6	0.3
	f4	34.78	0.01	145.9	0.1
	f4a	34.89	0.01	94.40	0.14
	f4b	35.10	0.01	273.4	0.2
	f5	69.44	0.01	121.6	0.1
	f6	102.8	0.01	123.1	0.1
	f7	135.9	0.1	133.8	0.1
SGSC	f1	8.743	0.024	1521.	3.
	f2	11.57	0.02	1177.	3.
	f3	17.56	0.02	779.3	1.5
	f4	35.01	0.03	404.4	0.5
	f4a	34.78	0.02	219.6	0.3
	f4b	35.18	0.02	800.7	1.9
	f5	69.46	0.02	245.1	0.1
	f6	102.6	0.1	202.9	0.1
	f7	135.6	0.1	196.6	0.1
MWC	f1	8.451	0.014	231.8	2.0
	f2	11.31	0.01	211.3	0.5
	f3	17.03	0.01	195.0	2.7
	f4	34.19	0.02	165.5	1.0
	f5	68.22	0.02	143.0	0.1
	f6	101.5	0.1	137.3	0.8
	f7	134.2	0.1	151.8	0.1
MLC	f1	8.439	0.006	200.2	0.9
	f2	11.24	0.01	158.2	0.6
	f3	16.78	0.03	125.7	2.6
	f4	33.45	0.01	105.3	0.1
	f5	66.32	0.02	122.8	0.1
	f6	98.80	0.01	146.5	0.1
	f7	130.9	0.1	174.7	0.1
MSC	f1	7.705	0.004	340.5	0.3
	f2	10.30	0.01	273.8	0.3
	f3	15.47	0.02	204.9	0.3
	f4	30.65	0.01	144.7	0.2

f5	61.34	0.01	116.0	0.1	
f6	91.00	0.01	114.9	0.1	
f7	120.8	0.1	138.4	1.0	

TRANSFORMER TYPE		Frequency (kHz)		Circuit Resistance (mΩ)	
		Average	Sample standard deviation	Average	Sample standard deviation
Calibration	f1	8.875	0.002	168.2	0.8
	f2	11.73	0.01	141.9	1.6
	f3	17.58	0.01	109.4	0.5
	f4	35.05	0.01	94.46	1.0
	f5	69.57	0.08	94.99	0.4
	f6	102.7	0.2	102.1	2.6
	f7	135.7	0.1	117.3	1.2
1SW	f1	8.842	0.004	173.9	0.4
	f2	11.67	0.01	148.3	0.2
	f3	17.45	0.01	115.4	0.3
	f4	34.55	0.01	106.8	0.4
	f5	67.74	0.02	111.5	0.2
	f6	98.54	0.01	134.8	0.2
	f7	129.5	0.1	158.8	0.2
$1/2$SW	f1	8.824	0.003	177.2	0.8
	f2	11.65	0.01	151.8	0.3
	f3	17.40	0.01	119.3	0.2
	f4	34.37	0.01	113.3	0.4
	f5	67.09	0.02	119.9	0.5
	f6	97.35	0.22	141.9	0.3
	f7	127.3	0.1	171.6	0.4
1WW	f1	7.560	0.004	475.9	1.1
	f2	9.663	0.003	436.8	0.6
	f3	13.60	0.01	465.0	1.2
	f4	23.26	0.04	649.5	1.3
	f5	37.12	0.04	1006.	9.
	f6	49.82	0.05	1156.	10.
	f7	59.93	0.11	1202.	5.

$^1/_2$SW$^1/_2$P	f1	8.260	0.037	288.0	0.4
	f2	10.77	0.01	278.6	1.8
	f3	15.53	0.01	278.8	0.4
	f4	28.50	0.01	396.9	0.3
	f5	49.33	0.04	609.4	1.9
	f6	68.24	0.03	658.3	0.7
	f7	84.12	0.16	810.7	2.9
$^1/_2$SW	f1	8.240	0.003	284.7	0.4
	f2	10.76	0.01	272.0	0.5
	f3	15.57	0.01	278.6	0.4
	f4	28.56	0.04	387.6	1.3
	f5	49.67	0.02	573.0	3.1
	f6	68.88	0.06	629.0	2.6
	f7	83.44	0.05	796.7	0.8
1LW	f1	7.646	0.005	498.2	1.9
	f2	9.760	0.007	475.5	0.8
	f3	13.77	0.02	443.4	0.9
	f4	23.42	0.01	473.2	0.2
	f5	37.52	0.04	570.9	3.9
	f6	48.26	0.02	649.5	0.7
	f7	57.36	0.01	732.5	2.3
1BWW	f1	8.476	0.005	261.5	2.6
	f2	11.10	0.01	239.4	0.9
	f3	16.19	0.01	233.3	1.7
	f4	30.19	0.01	308.4	0.3
	f5	54.98	0.02	433.9	1.9
	f6	76.18	0.13	515.3	3.3
	f7	94.78	0.04	581.9	0.3

A.3 MEASUREMENT OF CIRCUIT RESIDUAL RESISTANCE AND COIL RESISTANCES

The predicted resistance is plotted against the measured resistance for three identical coils made of copper, brass and german silver. A straight line is fitted to the three points and the residual resistance is then given by the value on the measured resistance axis where the straight line intercepts.

The procedure for calculating the predicted resistance values for the coils is as follows:

- The resistance is composed of two parts, namely the coil and connection lead resistance:

$$R_{prediction} = R_{coil} + R_{lead} \qquad (A.4)$$

- The conductivity values as used for the different conductive materials are:

Copper wire	$\equiv 5.8 \times 10^7$ S/m
Copper strip	$\equiv 5.65 \times 10^7$ S/m
Brass	$\equiv 1.52 \times 10^7$ S/m
German silver	$\equiv 4.35 \times 10^6$ S/m

- The predicted and measured (on a resistance bridge) dc resistance values have been set equal. Small deviations due to inaccuracies of dimensions, have been taken up by allowing the lead lengths to deviate slightly from the true lengths. These connection lead lengths, with the true values in brackets, are as follows:

SCC - 33 mm (32 mm)

SBC - 37 mm (32 mm)

SGSC - 32 mm (32 mm)

MWC - 24 mm (30mm)

- In the case of the strip conductor and litz wire coils the lead resistances could not be calculated because the structural impedance program (PIP) does not accommodate the specific configurations. Another factor to take into account is the difference between the length of the strand, and the effective litz wire length due to the twisting of the wire bundle to transpose the strand positions. Consequently, for these coils multiplication factors have been introduced, which are as follows:

MLC - 1.039

MSC - 1.017

Frequency (kHz)		$R_{residue}$ (mΩ)	Slope	Coefficient of Determination
f1:	8.88	35.11	1.008	1.000
f2:	11.8	27.41	1.038	1.000
f3:	17.6	26.71	1.017	1.000
f4:	35.0	37.33	0.979	1.000
f4a:	34.9	38.90	0.965	1.000
f4b:	35.1	42.00	1.013	1.000
f5:	69.6	48.51	1.002	0.999
f6:	103	62.11	1.014	1.000
f7:	136	81.21	1.033	0.999

The slope, being within 4% of unity, and the excellent coefficient of determination support the validity of the calibration of residual resistance values. The measurements at f4, f4a and f4b represent a four times variation in current and capacitor resistance. Under these conditions the residual resistance varies by a little more than ten percent , which indicates that the residual resistance is less dependent on the capacitor configuration and current level than on frequency variation. To simplify matters, the approximation is made that the residual resistance is only a function of frequency. To accommodate the variation in frequency during measurements on the multilayer coils due to differences in inductance values, the residual resistance at other frequencies are obtained from logarithmic interpolation between the values of the table on the previous page. The measured coil resistances are obtained by taking the difference between the observed total circuit resistances and the residual resistances.

COIL RESISTANCE				
Type	Frequency(kHz)	Theoretical(mΩ)	Measured(mΩ)	Error(%)
SCS	8.88	32.03	32.60	-2
	11.8	34.77	36.77	-5
	17.6	40.57	41.79	-3
	35.1	56.09	57.78	-3
	69.8	78.53	82.50	-5
	103	94.85	97.86	-3
	136	108.57	116.36	-7
SBC	8.88	94.85	95.25	-1
	11.8	96.09	98.60	-3
	17.6	99.42	100.45	-1
	35.0	114.57	108.57	+6
	69.4	150.83	146.18	+3
	103	182.54	182.97	0
	136	209.38	210.36	-5
SGSC	8.74	368.40	371.47	-8
	11.6	369.06	383.20	-4
	17.6	369.92	376.30	-2
	35.0	374.46	367.07	+1
	69.5	391.30	393.18	-1
	103	415.74	422.37	-2
	136	444.98	461.56	-4

MWC	8.45	42.86	49.17	-13
	11.3	55.58	61.30	-9
	17.0	82.84	84.15	-2
	34.2	148.03	128.17	+16
	68.2	222.68	188.98	+18
	102	277.97	225.27	+23
	134	323.21	282.36	+15
MLC	8.44	37.05	41.27	-10
	11.2	38.53	43.60	-12
	16.8	42.78	49.50	-14
	33.5	65.02	67.97	-4
	66.3	144.2	148.58	-3
	98.9	252.9	253.17	0
	131	371.9	373.96	-1
MSC	7.71	63.69	76.35	-17
	10.3	64.20	82.13	-22
	15.5	65.48	89.10	-27
	30.7	71.33	107.37	-34
	61.3	91.24	134.98	-32
	91.0	117.43	158.37	-25
	121	146.41	228.7	-36

A.4 MEASUREMENT OF TRANSFORMER RESISTANCE

The transformer ac resistance measurements at each frequency point were preceded by calibration measurements that entail measurement of circuit resistance with the transformer excluded from the circuit. The transformer ac resistance is obtained from the difference between the measured circuit resistances with and without the transformer included in the circuit. A complication encountered with this method is the shift in frequency when the transformer is connected, due to its leakage inductance. Another complexity, which is not present in the coil measurements, is the fact that we are dealing with primary referred resistance. The procedure devised to extract the most information from the experimental measurements is as follows:

- The same procedure as for the coils has been followed to calculate the connection lead resistance. Inaccuracies of dimensions have also been taken up by varying the connection lead lengths, which are (with the true values in brackets):

Secondary winding : 19 mm (20 mm)

1SW : 30 mm (25 mm)
$1/2$SW : 42 mm (42 mm)

1WW : 75 mm (80 mm)
$1/2$WW$1/2$P : 90 mm (40 mm)
$1/2$WW : 90 mm (40 mm)

1LW : 105 mm (80 mm)
1BWW : 95 mm (40 mm)

- The calibration resistance ($R_{calibration}$), allowing for the frequency shift, which is subtracted from the circuit resistance with transformers included, is deduced from the calibration measurement with the transformer removed ($R_{cal-observed}$) and residual resistance ($R_{residue}$) as follows;

$$R_{calibration} = K \ (R_{cal-observed}(f_c) - R_{residue}(f_c)) + R_{residue}(f_t) \qquad (A.5)$$

$$where \ \ K = \frac{R_{coils}(f_t)}{R_{coils}(f_c)}$$

with $R_{coils} \equiv$ the ac resistance of the serie/parallel coil configuration in,

the calibration circuit

$f_t \equiv$ frequency of circuit with transformer included,

$f_c \equiv$ frequency of circuit with transformer excluded.

TRANSFORMER RESISTANCE (mΩ)				
Type Frequency(kHz)	Calibration	Measured	Theoretical	Error(%)
1SW	8.84	168.2	5.7	4.15	-20
	11.7	141.9	6.4	4.80	-25
	17.5	109.4	6.0	5.54	-8
	34.6	94.46	12.3	9.31	-25
	67.7	94.99	16.5	22.64	+37
	98.5	102.1	32.7	40.38	+23
	130	117.3	41.5	61.52	+48
$1/2$SW	8.82	168.2	9.00	6.35	-29
	11.7	141.9	9.90	6.63	-33
	17.4	109.4	9.90	7.36	-26
	34.4	94.46	18.84	10.85	-42
	67.1	94.99	24.91	22.70	-9

	97.4	102.1	39.80	38.14	-4
	127	117.3	54.30	55.60	+2
1WW	7.6	168.2	307.7	245	-20
	9.7	136.6	300.2	280	-7
	13.6	103.6	361.4	364	+1
	23.3	75.2	574.3	651	+13
	37.1	71.0	936.8	1151	+23
	49.8	69.2	1086.8	1601	+47
	59.9	52.8	1149.2	1932	+68
$1/2$WW$1/2$P	8.3	168.2	119.8	108	-10
	10.8	140.8	137.8	136	-1
	15.5	106.9	179.8	204	+13
	28.5	85.6	311.3	461	+48
	49.3	80.1	529.3	901	+70
	68.2	75.6	582.7	1232	+110
	84.1	84.7	725.3	1473	+103
$1/2$WW	8.24	168.2	116.5	85.5	-27
	10.8	140.8	131.2	99.4	-24
	15.6	106.9	171.7	134	-22
	28.6	85.6	301.4	260	-14
	49.7	80.1	492.9	487	-1
	68.9	75.6	553.4	668	+21
	83.4	84.7	712.0	792	+11
1LW	7.65	168.2	330.0	290	-12
	9.76	136.6	338.9	295	-14
	13.8	103.6	339.8	308	-9
	23.4	75.2	398.0	358	-10
	37.5	71.0	499.9	475	-5
	48.3	69.2	580.3	597	+3
	57.4	52.8	679.7	723	+6
1BWW	8.48	168.2	93.3	84.1	-10
	11.1	141.9	97.5	96.7	-1
	16.2	108.8	124.5	129	+4
	30.2	86.2	222.2	249	+12
	55.0	83.6	350.3	475	+36
	76.2	86.8	428.5	645	+51
	94.8	89.9	492.0	778	+58

A.5 COMPUTER PROGRAM FOR FITTING THE ANALYTICAL FUNCTION TO THE MEASURED WAVEFORM, AND CALCULATING THE RESISTANCE AND INDUCTANCE

The voltage across the capacitor(s) is measured with a Tektronix 468 oscilloscope. The waveform is digitised into 512 points on the time axis; each point being a 8-bit number. The waveform is then fitted to the "best" analytical function as portrayed by equation (A.1). The fitting is done as follows:

- The period and phase is obtained through the average time interval between the zero crossings of the measured waveform.

- The damping constant is obtained by a numerical search for the minimum quadratic error between the analytical function and the measured waveform.

The input of data and execution of the program is done through a menu. To verify the accuracy of the fitted function, the program provides a graphic output as shown in Figure A.3.

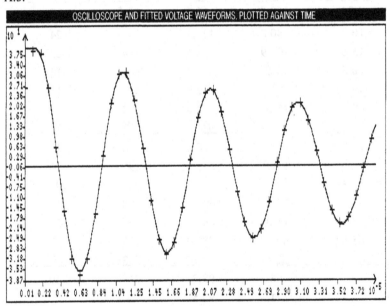

Figure A.3   An example of the measured waveform (solid line) and fitted function(crosses).

APPENDIX B

COMPUTER PROGRAM PIP
(Parasitic Impedance Program)

B.1 DESCRIPTION

PIP is a computer program which calculates the "parasitic" resistance, capacitance and inductance of a conductor and its return path. The resistance and internal inductance are calculated as function of frequency, while the capacitance and external inductance are frequency independent. The algorithm was developed, specifically having connection structures of power converters in mind. The program is capable of analysing the five different structures shown in Figure B.2. With these it should be possible to cover most connections in power converter circuits. Something to be kept in mind during a modelling exercise is that the high frequency current components in a conductor that runs closely to a conductive plane, normally a heatsink, will induce currents, equal but opposite in direction, in the conductive plane. This can be ascribed to the fact that alternating currents tend to rearrange themselves in conductive structures to give minimum inductance in current loops.

Another convenient feature of the program is that it permits analysis of two dielectic material configurations as shown in Figure B.1. The dielectric slab configuration is particularly useful in cases such as double sided printed circuit boards, having a conductive plane on the one side, or a conductor positioned on and isolated from a heatsink with isolation tape. The uniform dielectric configuration is for example, applicable to conductors which are suspeded in air or immersed in oil.

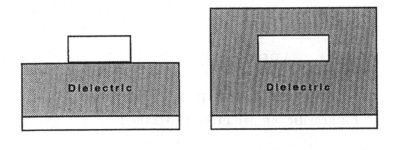

Slab   **Homogeneous**

Figure B.1 Dielectric configurations.

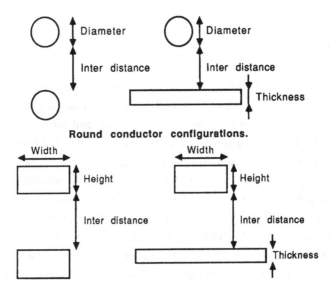

Round conductor configurations.

Rectangular conductor configurations.

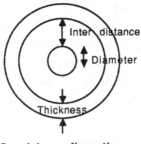

Coaxial configuration.

Figure B.2 Conductor configurations.

B.2 <u>MENUS</u>

B.2.1 <u>Main menu</u>

When the program is initially executed, the main menu is displayed, as shown in Figure B.3. The user is faced with a selection of possible configurations, which include wire, rectangular conductors, coaxial configurations as well as a conductive plane such as a heatsink. By typing the number of the selected configuration, program branches to the particular sub-menu. The user is also given a choice between a "frequency sweep" of ac impedance between 10Hz and 100MHz or the calculation of ac impedance at selected frequencies.

```
ANALYSIS OF PARASITIC IMPEDANCES - VERSION 1
                COPYRIGHT - SNO-WNNR, RAU, 1986
                Author - Braham Ferreira


MAIN MENU


  1. Twin Round Conductors.
  2. Round Conductor and Conductive Plane.
  3. Coaxial Conductors.
  4. Twin Rectangular Conductors.
  5. Rectangular Conductor and Conductive Plane.

    Single frequency

    Frequency Spectrum <==

Terminate Program.
```

Figure B.3 Main menu of PIP.

B.2.2 <u>Sub-Menu</u>

A sub-menu (Figure B.4) can be divided into two parts, namely, the geometrical parameters and output control. The geometrical parameters entail dimensioning of the configuration and selection of conductive materials and dielectric. Two types of dielectric spatial configuration can be analysed, namely, a homogeneous distribution or a dielectric slab between the two conductors, as illustrated in Figure B.1. The submenus give the user a choice between graphic and tabular output format. (See Figures B.5 and B.6.) When the tabular format is selected, the results are also written to one of nine output files; **IMPn.DAT** with **n** = 1 to 9.

```
ANALYSIS OF PARASITIC IMPEDANCES - VERSION 1

SUB-MENU FOR
A ROUND CONDUCTOR AND A CONDUCTIVE PLANE

    Diameter of conductor = 0.0100 m
    Inter distance plane and conductor = 1.0000 m
    Thickness of conductive plane = 0.1000 m
    Length of conductor = 1.0000 m
    Open ends (Number = 0,1 or 2) = 0
    Conductor material(Copper,Brass,Aluminium,Own value) = Copper
    Material of conductive plane(Copper,Brass,Aluminium,Own value) = Copper
    Dielectric material:
        Uniform or Slab - U
        Permittivity (relative value) =   1.0


Result;  X - Graphical
         Y - Tabular and Datafile    <===
Number of datafile (1 tot 9)
    Name of file: IMP1.DAT

Go to main menu.
```

Figure B.4 An example of a sub-menu.

```
IMPEDANCE OF GIVEN CONFIGURATION.

Capacitance(pF) = 9.55E+000
```

Frequency	Resistance(m-ohm)	Inductance(nH)
1.00E+000	2.195E-001	1.220E+003
3.16E+000	2.196E-001	1.217E+003
1.00E+001	2.197E-001	1.216E+003
3.16E+001	2.199E-001	1.215E+003
1.00E+002	2.216E-001	1.214E+003
3.16E+002	2.352E-001	1.212E+003
1.00E+003	3.210E-001	1.203E+003
3.16E+003	5.314E-001	1.187E+003
1.00E+004	8.963E-001	1.177E+003
3.16E+004	1.548E+000	1.171E+003
1.00E+005	2.708E+000	1.168E+003
3.16E+005	4.771E+000	1.166E+003
1.00E+006	8.441E+000	1.165E+003
3.16E+006	1.497E+001	1.164E+003
1.00E+007	2.657E+001	1.164E+003
3.16E+007	4.721E+001	1.164E+003
1.00E+008	8.391E+001	1.163E+003

```
Press any key to return to the menu.
```

Figure B.5 An example of the result in tabular form

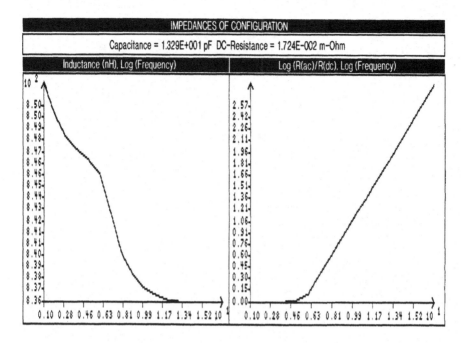

Figure B.6 An example of the result in graphical form.

APPENDIX C

COMPUTER PROGRAM TID
(Transformer and Inductor Design)

C.1 DESCRIPTION

TID is a general computer program to calculate losses in windings wound on magnetic cores. The program originated to fulfill the need to analyse and tune the magnetic components in the circuit for efficiency and ease of manufacture. To achieve these objectives the program provides four analytical approaches:

a) Primary Referred Resistance

The losses are converted to a frequency dependent resistance which can be incorporated in a circuit analysis of the converter.

b) Losses in the Windings

The total losses per winding section for the specified current waveforms are calculated to enable the designer to make an assessment of the efficiency of the magnetic component and its contribution to the overall system efficiency.

c) Distribution of Losses in Winding Sections

By displaying the distribution of stray losses inside windings, the designer is able to locate the areas and winding sections where the dissipation is the largest, which is a valueble tool for loss optimisation and supplies essential information for the design of heat removal schemes.

d) Magnetic Stray Field in the Window Area

The intensity and distribution of the magnetic stray field inside the winding window gives an indication of the energy flow between and incidence of proximity losses inside the different winding sections of the magnetic component. Methods of calculating the magnetic fields are discussed in the next chapter.

TID does a two dimensional analysis on winding sections contained within a winding window. It is specifically geared to optimise magnetic transducers at high frequencies. Virtually any kind of magnetic core (E-, U- or Pot cores), with or without airgaps, can be analysed with this program. TID has a few limitations due to the complexity of the electromagnetic equations involved. The algorithm neglects the effect of the cylindrical curvature on the magnetic field distribution inside the window. Provided that the curvature is small in relation to the radial height of the windings, and the axial length to diameter ratio of winding sections is not much larger than one, the two dimensional analysis should be sufficiently accurate. Another limitation is that fringing of the magnetic field around airgaps is not accounted for due to limitations of the method of images being used to calculate the fields. The program therefore, is not able to analyse localised heating around airgaps, but instead assumes the heating caused by an airgap to be uniformly distributed along the radial height of the winding adjacent to the air gap. In order to speed up the calculation procedures and to keep computer memory requirements down, the algorithm assumes the current distribution of winding sections to be uniform. The consequence of this is that the windings inside the window should have enough turns so that the individual contribution of each turn to the total magnetic field at a point is small.

Bearing above considerations in mind it, it should be possible to calculate within 10 or 20% accuracy the winding losses together with their distribution for most practical magnetic transducer configurations. This program is ideally suited to compare alternative windings configurations in so far as their losses are concerned, since the relative loss figures will be more accurate than the total loss figures.

C.2 STRUCTURAL CONFIGURATION

The program does essentially a two dimensional analysis on the windings of a magnetic transducer. The user specifies the dimensions of the winding window, i.e. the boundary between the magnetic core and the space in which the windings are wound. The configuration parameters can subsequently be subdivided into two catagories, namely magnetic core and winding sections.

C.2.1 Winding configuration

TID allows the user to place winding sections arbitrary inside the winding window, provided they do not overlap. The program differenciates between primary and secondary windings according to the sign of the current coefficients; positive for primary and negative for secondary. TID allows windings to be subdivided into sections placed at different positions inside the winding window, all of which are connected in series. No limitation is placed on the amount of secondary windings and sections thereof, provided the maximum number of winding sections, allowed by the program version, is not exceeded. On the primary side, however, only one winding is allowed (which can be subdivided into sections).

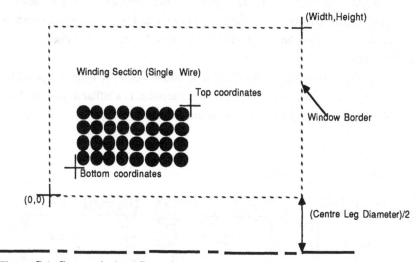

Figure C.1 Geometrical configuration.

For the purpose of communication between user and computer, the window area is considered to be a XY-cartesian coordinate plane with the top left hand corner of the computer screen being the zero reference point, and the x-axis being parallel and the y=0 line being closest to the winding axis. As indicated in Figure C.1 a winding section is designated to the window plane by entering its top and bottom coordinate points.

TID permits the following conductor configurations:
  a) Single Wire.

> The windings can be solid round wire. The diameter of the conductive material must be specified. This diameter would, therefore, exclude the dimensions of any

isolation sheathing present on the wire, and the user has to allow for it in the dimensions of the particular winding section.

b) Litz Wire.

The windings can be fully transposed litz wire. The diameter of the complete litz bundle, diameter of the individual strands as well as the number of filaments must be specified. As in the case of the single wire, the strand diameter only specifies the diameter of the conductive material, and space must be allowed for isolation in the litz bundle diameter.

c) Strip Conductor.

The windings can be strip conductors of which the width is much larger than the thickness. Only the thickness of the strip need to be specified, since TID assumes the width of the strip to be the same as that of the winding section. The thickness of the strip includes only the conductive material, and provision must be made in the choice of the winding section height for any isolation material present.

d) Bifilar Windings.

TID permits centre tapped windings to be bifilar wound. It assumes that such a winding consists of two winding sets wound together in a bifilar fashion, and that only one of the winding sets carries current at a time.

C.2.2 Magnetic core configuration

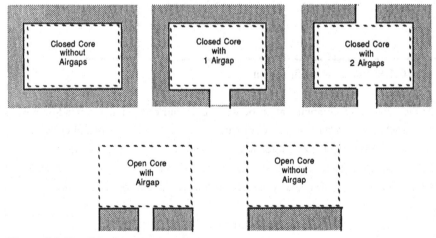

Figure C.2 Possible core arrangements.

TID is only concerned with characteristics of the part of the core that borders on the winding window (it does not do a magnetic circuit analysis of the core) and neglects curvature of the core. The user has the option to include one or two air gaps in the configuration. He can also specify the permeability of the magnetic material, but it must be kept in mind that the program does not take into account the contribution of the other half of the winding, when calculating the magnetic field distribution. This does not normally encroach on the validity of the field solution, provided that the permeability is suficiently high. (Program, CID, is available for the analysis of air cored inductors.) TID allows the user also to define open and closed core configurations, each spanning a portion of the circumference of the windings, thus enabling him to determine the characteristics of coils wound on E- and U+I-cores, which have for one part of the turn a magnetic enclosed winding window, while the other part the window is "open", i.e. only one window boundary interfaces with magnetic material.

Virtually all types of cores can be analysed which include: E-cores, U+I-cores, Pot cores, rods. Approximate analysis of machine slots and toroides can also be done. If analysis of a machine slot is attempted it must be borne in mind that the program assumes the coils to be cylindrical shaped, and the dimensions should be accordingly adapted. With a toroid transformer a closed core configuration with one airgap which extends over the total winding width has to be selected, and the permeability must be set to a high value. The field distribution of the toroid (ignoring curvature) is the same as for such a two dimensional window, with the toroid circumference contained within the width of the window. However, the error caused by ignoring the curvature can be substantial if the experimental results of Chapter 8 on long solenoids are to be applied on many toroid configurations. A more accurate program for toroids using circular current filaments is under developement, but is not included as part of this book.

C.3 MENU DESCRIPTION

C.3.1 Main Menu

When the main program is initially executed, the main menu is displayed on the screen. (See Figure C.3) From this menu the user gains access to two sub-menus for configuration parameters. The user is given a choice of four output formats, namely total power dissipation in each of the winding sections and the whole winding configuration, a power dissipation density plot of one of the winding sections, the effective resistance as a function of frequency due to copper losses, and finally the facility to make a plot of the magnetic field distribution inside the winding window. The data of the complete configuration can be read from or written to any of 9 datafiles, labelled **KONFIGn.DAT** with **n** being a number between 1 and 9.

```
TRANSFORMER AND INDUCTOR WINDING DESIGN - VERSION 1
                         COPYRIGHT - SNO-WNNR, RAU, 1986
                            Author - Braham Ferreira
MAIN MENU

Sub-Menus:
   X. General Configuration.
   Y. Winding Configuration.

Output Control:
     Total Power Dissipation
     Effective Resistance                    <===
     Power Dissipation Density Plot
     Magnetic Field Intensity Plot

Store in/Read from datafile(1 to 9):KONFIG3.DAT

Calculate results

Quit Program.
```

Figure C.3 Main menu example.

C.3.2 Output of results

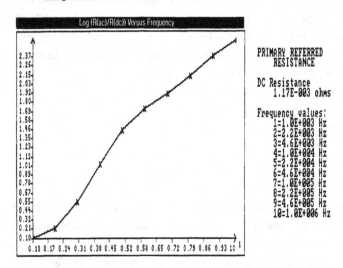

Figure C.4 Graph of effective resistance example.

TOTAL POWER DISSIPATION.

```
LOSSES:
      Section(1) = 9.32E-001 watt
      Section(2) = 1.93E-001 watt
      Section(3) = 5.62E-001 watt
                   --------------
         Total = 1.69E+000 watt
```

Figure C.5. Table of power dissipation in winding sections example.

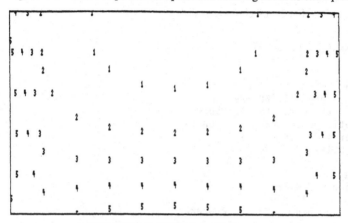

POWER DENSITY OF SECTION 1 (watt/m^3)

1=1.4E+005   2=3.2E+005   3=5.1E+005   4=7.0E+005   5=8.8E+005

Figure C.6 Power dissipation distribution inside a winding section example.

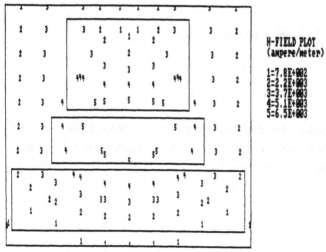

H-FIELD PLOT
(ampere/meter)

1=7.8E+002
2=2.2E+003
3=3.7E+003
4=5.1E+003
5=6.5E+003

Figure C.7 Magnetic field distribution inside winding window example.

Figures C.4 to C.7 give examples of the output formats. In case the resistance values are needed for circuit analysis by other programs, TID also permits the user to store the resistance values in separate data files called **RESISTANCEn.DAT** with n from 1 to 5.

C.3.3 General configuration menu

Within this menu the user can configure the magnetic core, the window dimensions and the number and value of the frequency components.

```
SUB-MENU FOR GENERAL CONFIGURATION.


Magnetic Core Parameters:
   Permeability of core = 1.00E+004
   Closed Window Ratio(persentage) = 50.00
   Open Window Ratio(persentage) = 50.00
   Airgaps (Number = 0,1 or 2) = 1
   Length of airgap(s) =0.00100

Window Dimensions:
   Height of Window = 0.01600m
   Width of Window = 0.04000m
   Diameter of Centre Leg of Core = 0.02000m

Frequency Components:
   First Harmonic = 1.00E+003
   Number of frequency components (maximum of 5) = 3

Go to main menu.
```
Figure C.8 General configuration sub-menu example.

C.3.4 Winding configuration menu

This menu displays the essential configuration parameters of up to six different winding sections. For each winding section a separate sub-sub-menu exists in which the winding parameters of that section can be entered or altered.

SUB-MENU FOR WINDING CONFIGURATION.

Number of winding sections = 3

Winding Sections
1: Bottom coordinates=(0.0100,0.0010) Top coordinates=(0.0300,0.0070)
    Turns=(1x5) Current=(4.2E+001,1.3E+001,7.8E+000)Amp
    Type: Copper,Monofilar,Strip(0.00100m,0.02000m)
2: Bottom coordinates=(0.0075,0.0075) Top coordinates=(0.0325,0.0105)
    Turns=(15x2) Current=(0.0E+000,0.0E+000,0.0E+000)Amp
    Type: Copper,Monofilar,Wire(0.00150m)
3: Bottom coordinates=(0.0010,0.0110) Top coordinates=(0.0390,0.0150)
    Turns=(25x3) Current=(-2.5E+000,-8.5E-001,-5.3E-001)Amp
    Type: Copper,Monofilar,Litz(0.00120m,0.00030m,10)

Go to main menu.

Figure C.9 Winding configuration sub-menu example

SUB-SUB-MENU FOR CONFIGURATION OF WINDING SECTION 1

Conductor type: X - Wire
                  Y - Litz Wire
                  Z - Strip Conductor   <===
Winding Technique: E - Monofilar   <===
                    F - Bifilar
Material of conductors(Copper,Brass,Aluminium,Own type) = Copper
Number of turns per layer = 1
Amount of layers = 5
Bottom corner (x,y) = (0.01000,0.00100)
Upper corner (x,y) = (0.03000,0.00700)
Thickness of strip = 0.00100
Width of strip = 0.02000
Coefficients of current amplitudes: 1st harmonic = 42.300 Amp
                                      3rd harmonic = 13.200 Amp
                                      5th harmonic =   7.800 Amp

Go to winding menu.

Figure C.10 Winding configuration sub-sub-menu example.

APPENDIX D

COMPUTER PROGRAM CID
(Coil Inductor Design)

D.1 <u>DESCRIPTION</u>

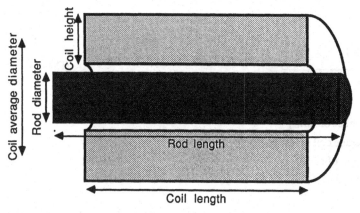

Figure D.1 Cross section through a coil.

CID is a program which supplements TID, and is specifically written to analyse air cored coils and inductors wound on a magnetic rod. The basic configuration of the coil, as analysed by the program, is shown in Figure D.1. The magnetic rod is optional, but if included, it has to be at least as long as the coil. The user is given a choice of outputs between frequency dependent resistance and power dissipation due to an arbitrary periodic current waveform. Two further options on resistance calculation are provided, namely, a repetitive resistance calculation at selected frequencies and the calculation of the resistance of the coil at the ground frequency and first eight harmonic Fourier components. During the calculation of the total losses in the coil, the user specifies the dc component and Fourier coefficients of the current flowing through the inductor. Another useful feature of CID is that it also calculates the inductance of the coil. The combined analysis of inductance and losses, permits one to optimise an inductor for its reactive power rating, losses and cost.

CID features two algorithms for the calculation of magnetic field intensity inside the windings. The first algorithm is essentially the same as the one being used in TID, and makes the approximation that the curvature of the coil turns can be neglected. Experimental work, however, indicated that the approximation is not valid for long solenoids and configurations where the curvature is comparable to the coil height. To this end, the second algorithm has been included which calculates the magnetic field of a circular current distribution. The penalty to be paid for this more accurate method is longer calculation times.

The following empirical equation given by reference [1] is used to calculate the inductance of a coil;

$$L = \frac{N^2 d^2}{1.01(0.45d + b + c + bc/(2d))} \qquad (D.1)$$

where N = number of turns,
         d = average coil diameter,
         b = length of coil,
         c = thickness of coil.

The feature to include a magnetic rod to enhance the inductance of the coil is incorporated in the program. The effective relative permeability is calculated by doing curve fitting to the graph in reference [2], page 41. If μ_r lies between 10 and 100 parameters a,b,c are calculated as follows:

$$c = -1 + 0.435 \ln \mu_r \qquad (D.2)$$

$$a = 3.54 + 12.46\,c \qquad (D.3)$$

$$b = 0.261 + 0.137\,c \qquad (D.4)$$

If μ_r is larger than 100 then:

$$c = -2 + 0.434 \ln \mu_r \tag{D.5}$$

$$a = 16 - 6.238\, c \tag{D.6}$$

$$b = 0.398 + 0.59\, c \tag{D.7}$$

Setting the lenght and diameter of the rod equal to l and d respectively, a value μ_{e1} is calculated from the expression:

$$\mu_{e1} = a\,(l/d)^b \tag{D.8}$$

Now calculate μ_{e2} from:

$$\mu_{e2} = 3.4\,(l/d)^{1.415} \tag{D.9}$$

The effective permeability of the rod is then the minimum of these two values;

$$\mu_e = \text{minimum}\,(\mu_{e1}, \mu_{e2}) \tag{D.10}$$

CID permits the following conductor types:
 a) Single Wire.

> The windings can be solid round wire. The diameter of the conductive material must be specified. This diameter would therefore exclude the dimensions of any isolation sheathing present on the wire, and the user has to allow for it in the dimensions of the particular winding section.

 b) Litz Wire.

> The windings can be fully transposed litz wire. The diameter of the complete litz bundle, diameter of the individual strands, as well as the number of filaments, must be specified. As in the case of the single wire, the strand diameter only specifies the diameter of the conductive material, and space must be allowed for isolation in the litz bundle diameter.

 c) Strip Conductor.

> The windings can be strip conductors of which the width is much larger than the thickness. The thickness and width of the strip includes only the conductive material, and provision must be made in the choice of the coil height and length for any isolation material present.

D.2 <u>MENU</u>

CID has a single menu divided into six sections, as shown in Figure D.2; conductor type, conductor dimensions, coil measurements, magnetic core, current components and frequencies, and calculation of results. Note that all the configuration information is displayed, but that the values can't be changed while the menu is in this mode. By repetitive pressing of the spacebar, different sections can be selected and opened to change the configuration type and parameter values. The selected section is indicated on the menu by an arrow, as can be seen on the menus of Figures D.3 and D.4.

The first section, "conductor type", when opened, also contains the frequency setting for resistance calculation and selection between cylindrical and rectangular coordinate system for magnetic field calculation (Figure D.3). Figure D.4 shows the menu of the section, "current components and frequencies", when the output mode is set to power dissipation calculation. To enter the coeffients of current amplitudes the user first has to conduct a Fourier analysis of the measured or predicted current waveforms to determine the current coefficients. Note that the Fourier coefficients are peak values and not rms values. Finally the menu displays after calculation of results are shown in Figures D.5 and D.6.

```
ANALYSIS OF COILS - Version 1 by Braham Ferreira
        COPYRIGHT - SNO CSIR, RAU, 1986
                          (Press spacebar to change values)
CONDUCTOR TYPE:
  Single wire, Copper
CONDUCTOR DIMENSIONS:
  Diameter = 0.00200m
COIL MEASUREMENTS:
  Turns/layer, Layers = 4, 2
  Length, Average dia.,Height = 0.0200m, 0.1000m, 0.0050m
MAGNETIC CORE:
  None
CURRENT COMPONENTS AND FREQUENCIES:
  Fundamental frequency = 1.00E+003
CALCULATE NEW RESULTS:       O - Power dissipation
                             P - Resistance values <===
End program.
```

Figure D.2. Menu after initialisation of the program.

```
ANALYSIS OF COILS - Version 1 by Braham Ferreira
        COPYRIGHT - SNO CSIR, RAU, 1986
                            (Press spacebar to change values)
CONDUCTOR TYPE:   <===
  G - Solid <==   H - Litz         I - Strip
  Material (Copper,Brass,Aluminium,Own value) = Copper
  Frequency setting? (Repetitive/Fourier) - FOURIER
  A:Cylindrical or B:Rectangular coordinates? - RECTANGULAR
CONDUCTOR DIMENSIONS:
  Diameter = 0.00200m
COIL MEASUREMENTS:
  Turns/layer, Layers = 4, 2
  Length, Average dia.,Height = 0.0200m, 0.1000m, 0.0050m
MAGNETIC CORE:
  None
CURRENT COMPONENTS AND FREQUENCIES:
  Fundamental frequency = 1.00E+003
CALCULATE NEW RESULTS:      O - Power dissipation
                            P - Resistance values <===
End program.
```

Figure D.3 Menu with "conductor type" section selected.

```
ANALYSIS OF COILS - Version 1 by Braham Ferreira
        COPYRIGHT - SNO CSIR, RAU, 1986
                            (Press spacebar to change values)
CONDUCTOR TYPE:
  Single wire, Copper
CONDUCTOR DIMENSIONS:
  Diameter = 0.00200m
COIL MEASUREMENTS:
  Turns/layer, Layers = 4, 2
  Length, Average dia.,Height = 0.0200m, 0.1000m, 0.0050m
MAGNETIC CORE:
  None
CURRENT COMPONENTS AND FREQUENCIES:  <===
  Fundamental frequency = 1.00E+003    Number of components (max 9) = 9
  Coefficients of current amplitudes:
  DC-component =   1.000 Amp   1-harmonic =   1.000 Amp
  2-harmonic =   0.500 Amp   3-harmonic =   0.200 Amp
  4-harmonic =   0.100 Amp   5-harmonic =   0.100 Amp
  6-harmonic =   0.100 Amp   7-harmonic =   0.100 Amp
  8-harmonic =   0.100 Amp   9-harmonic =   0.100 Amp
CALCULATE NEW RESULTS:      O - Power dissipation  <===
                            P - Resistance values
End program.
```

Figure D.4 Menu with "current components and frequencies" section selected.

```
ANALYSIS OF COILS - Version 1 by Braham Ferreira
          COPYRIGHT - SNO CSIR, RAU, 1986
                              (Press spacebar to change values)
CONDUCTOR TYPE:
  Single wire, Copper
CONDUCTOR DIMENSIONS:
  Diameter = 0.00200m
COIL MEASUREMENTS:
  Turns/layer, Layers = 4, 2
  Length, Average dia.,Height = 0.0200m, 0.1000m, 0.0050m
MAGNETIC CORE:
  None
CURRENT COMPONENTS AND FREQUENCIES:
  Fundamental frequency = 1.00E+003
  Coefficients of current amplitudes = (1.0E+000,1.0E+000
  ,5.0E-001,2.0E-001,1.0E-001,1.0E-001,1.0E-001,1.0E-001,1.0E-001,1.0E-001)
CALCULATE NEW RESULTS:     O - Power dissipation  <===
                           P - Resistance values
  Inductance of coil =     8.99uH
  Power dissiption of coil =   0.0183Watt
End program.
```

Figure D.5 Menu displaying calculated power dissipation.

```
ANALYSIS OF COILS - Version 1 by Braham Ferreira
          COPYRIGHT - SNO CSIR, RAU, 1986
                              (Press spacebar to change values)
CONDUCTOR TYPE:
  Single wire, Copper
CONDUCTOR DIMENSIONS:
  Diameter = 0.00200m
COIL MEASUREMENTS:
  Turns/layer, Layers = 4, 2
  Length, Average dia.,Height = 0.0200m, 0.1000m, 0.0050m
MAGNETIC CORE:
  None
CURRENT COMPONENTS AND FREQUENCIES:
  Fundamental frequency = 1.00E+003
CALCULATE NEW RESULTS:     O - Power dissipation
                           P - Resistance values <===
  Inductance of coil =     8.99uH
  Resistance at;
    1.00E+003Hz =  0.0139 Ohm    2.00E+003Hz =  0.0144 Ohm
    3.00E+003Hz =  0.0151 Ohm    4.00E+003Hz =  0.0160 Ohm
    5.00E+003Hz =  0.0172 Ohm    6.00E+003Hz =  0.0184 Ohm
    7.00E+003Hz =  0.0197 Ohm    8.00E+003Hz =  0.0211 Ohm
    9.00E+003Hz =  0.0225 Ohm
End program.
```

Figure D.6 Menu displaying calculated resistance values.

D.3 <u>REFERENCES</u>

[1] E. Loefgren; "Naeherungsformelu zur Berechnung der Induktivitaet kreisformiger Spulen", Elektrotechnische Zeitschrift 71, Jahrgng Heft 6,1950, pp. 148-9.

[2] MF DeMaw; "Ferromagnetic-core design and application handbook"; Prentice-Hall, Inc -1981.

LIST OF SYMBOLS

a	constant
A	magnetic vector potential
α	complex inverse of skin depth
b	constant
B	magnetic flux density, susceptance
c	speed of light, constant
C	capacitance, constant
d	diameter
D	electric flux density
δ	skin depth
E	electric field intensity
ε	permittivity
f	frequency
F	skin effect geometrical factor
g	constant
G	proximity effect geometrical factor, conductance
h	thickness (height) of strip
H	magnetic field intensity
I	current
J	current density
k	constant
l	lenght
L	inductance
λ	ratio of height or diameter/$\sqrt{2}$ to skin depth
p	packing factor
r	radius
P	power
R	resistance
S	Poynting vector, complex power, mechanical stress
σ	conductivity
μ	permeability

v	velocity
V	electric potential
ν	ratio of height to skin depth
φ	electric scalar potential
Φ	magnetic flux
w	width of strip
ω	radial frequency
Ω	magnetic scalar potential
X	reactance
Z	impedance
ζ	√2 times inverse of skin depth

INDEX

A

AC energy flux   23
AC inductance   4,7,62,63,96
AC resistance   4,62,63,75
Air gap   100
Analytical method   55

C

Calometric measurement   107
Capacitance   10, 28, 113
Capacitor   24
Circuit network   42
Coil   (see also inductor)   6, 108, 113
Complex power   20,39
Conductive plane   75
Connection structure (see also power line)   41-48, 67-81
Cylindrical (round) conductor   62, 63, 85

E

Eddy currents   49-65
Effective impedance   42
Efficiency   30, 38
Electric machine   31
Energy
      flow, flux   (see Poynting vector)
      storage   17, 25, 26
Equivalent impedance   41
Experimental measurements   78, 108, 113, 129-144
External impedance   70

Power conditioning   35-48
Power line   19
Poynting vector   15-38, 125
Proximity effect   1, 73, 83-96

R

Reactive power   40, 45
Reflected energy flux   24
Resonant measurement   108, 113, 127, 129

S

Single layer coil   67, 108
Skin effect   2, 73, 83-96
Snubber   45
Stranded wire (litz)   88
Stray reactance   21, 45
Strip conductor   56, 58, 59, 85, 90
Structural impedance   67-81, 126
Switchmode circuits   36, 44

T

Termination of power lines   24
Transformers   10, 28, 113
Transmission line   18

W

Wattmeter measurement   107
Winding section   93